AS/A-LEVEL YEARS 1 AND 2

STUD[illegible]

EDEXCEL

Biology B

Practical assessment

Dan Foulder

The author and publisher would like to thank Richard Fosbery for his assistance with the preparation of this book.

Hodder Education, an Hachette UK company, Blenheim Court, George Street, Banbury, Oxfordshire OX16 5BH

Orders

Bookpoint Ltd, 130 Park Drive, Milton Park, Abingdon, Oxfordshire OX14 4SE

tel: 01235 827827

fax: 01235 400401

e-mail: education@bookpoint.co.uk

Lines are open 9.00 a.m.–5.00 p.m., Monday to Saturday, with a 24-hour message answering service. You can also order through the Hodder Education website: www.hoddereducation.co.uk

ISBN 978-1-4718-8516-7

First printed 2017

Impression number 5 4 3 2 1

Year 2021 2020 2019 2018 2017

This guide has been written specifically to support students preparing for the Edexcel AS and A-level Biology examinations. The content has been neither approved nor endorsed by Edexcel and remains the sole responsibility of the author.

Cover photo: Elena Pankova/Fotolia; p. 26, Ian Couchman, Cambridge International Examinations; p. 77, J.C. Revy, ISM/Science Photo Library

Typeset by Integra Software Services Pvt. Ltd, Pondicherry, India

Printed in Slovenia

Contents

Skills Guidance

Questions & Answers

About this book

This student guide is intended to help you develop the skills in practical biology required as part of Edexcel's A-level Biology B specification. The guide covers all the required A-level core practicals and the skills you are required to develop as you complete your practical work.

These skills are assessed in two ways:

- in the AS and A-level examination papers
- in the A-level practical endorsement

Some of the questions in the two AS exam papers and the three A-level exam papers will assess your understanding of practical skills and your ability to apply them to familiar and unfamiliar contexts. Some of the planning and implementation skills cannot be assessed in a written examination, so will be assessed by your teachers while you are carrying out practicals during your course.

During your course you will practise many mathematical skills. This guide covers the mathematical skills required for the core practicals.

This guide is divided into two sections:

Skills Guidance begins with a brief guide to some of the general skills and procedures that will be used across the core practical activities. This is followed by detailed guidance on each of the core practicals, including the biology involved, common mistakes and ideas for further work. The core practicals do not have to be carried out in the way detailed in this guide, but these are common examples that allow you to achieve the relevant Common Practical Assessment Criteria (CPAC). They also prepare you fully for possible exam paper questions on the core practicals. Required maths skills are also covered in the core practicals where appropriate, although not all the required maths skills from across the whole specification are included in this guide.

The **Questions & Answers** section contains examples of some of the types of question set in A-level Paper 3: General and Practical Principles in Biology, together with answers written by two students, one of whom makes many errors. There are comments on all the answers. These comments are intended to guide you to write concise answers that show understanding of the key skills in practical work in biology.

If you try all of the practical exercises and the questions in the Questions & Answers section before looking at the answers, you will begin to think for yourself and develop the necessary techniques for answering exam questions and performing well in your practical work in the lab and in the field.

Hazard and risk

All practical tasks described in this guide should be risk assessed by a qualified teacher before being performed either as a demonstration or as a class practical. Safety goggles and a laboratory coat or apron must be worn where it is appropriate to do so. The author and the publisher cannot accept responsibility for safety.

How the practical element is assessed

The Edexcel Biology specification requires you to experience teaching and learning opportunities that involve regular practical work.

If you are studying AS biology then you need to have studied core practicals 1–8 covered in this guide. These practicals will be assessed through the written examinations. This means you could be asked questions on them in either of the two written exam papers. There is no practical endorsement (see below) for the AS qualification, so you will not be assessed by your teacher against the CPAC criteria on page 8.

In the A-level course you will complete a minimum of 16 core practical activities. These will then contribute to the practical endorsement element of the A-level biology qualification. The practical endorsement is graded as pass or fail. In order to pass you need to show evidence of all the required skills (shown below). These will be assessed by your teacher as you complete the practicals.

The skills, knowledge and understanding you develop will also be assessed in written examinations in the context of the required practical activities and other practicals. The skills you need to develop can be split into those that can be assessed through written examinations and those that will be assessed by teachers through the practical activities.

Practical skills assessed in written examinations

Independent thinking

- Solve problems set in practical contexts.
- Apply scientific knowledge to practical contexts.

Use and application of scientific methods and practices

- Comment on experimental design and evaluate scientific methods.
- Present data in appropriate ways.
- Evaluate results and draw conclusions with reference to measurement uncertainties and errors.
- Identify variables including those that must be controlled.

Numeracy and the application of mathematical concepts in a practical context

- Plot and interpret graphs.
- Process and analyse data using appropriate mathematical skills, as exemplified in the mathematical appendix for each science.
- Consider margins of error, accuracy and precision of data.

Instruments and equipment

- Know and understand how to use a wide range of experimental and practical instruments, equipment and techniques appropriate to the knowledge and understanding included in the specification.

Skills assessed by your teacher while you carry out practicals

Independent thinking

- Apply investigative approaches and methods to practical work.

Use and apply scientific methods and practices

- Safely and correctly use a range of practical equipment and materials.
- Follow written instructions.
- Make and record observations.
- Keep appropriate records of experimental activities.
- Present information and data in a scientific way.
- Use appropriate software and tools to process data, carry out research and report findings.

Research and referencing

- Use online and offline research skills, including websites, textbooks and other printed scientific sources of information.
- Correctly cite sources of information.

Instruments and equipment

- Use a wide range of experimental and practical instruments, equipment and techniques appropriate to the knowledge and understanding included in the specification.

You also need to be able to show evidence of competence in the 12 practical techniques listed below. These differ from the more general practical skills above as they relate to specific experimental techniques.

1. Use appropriate apparatus to record a range of quantitative measurements (to include mass, time, volume, temperature, length and pH).
2. Use appropriate instrumentation to record quantitative measurements, such as a colorimeter or potometer.
3. Use laboratory glassware apparatus for a variety of experimental techniques, to include serial dilutions.
4. Use of a light microscope at high power and low power, including use of a graticule.
5. Produce scientific drawing from observation, with annotations.
6. Use qualitative reagents to identify biological molecules.

7 Separate biological compounds using thin-layer/paper chromatography or electrophoresis.
8 Safely and ethically use organisms to measure:
   - plant or animal responses
   - physiological functions
9 Use microbiological aseptic techniques, including the use of agar plates and broth.
10 Safely use instruments for dissection of an animal organ, or plant organ.
11 Use sampling techniques in fieldwork.
12 Use ICT such as computer modelling, or data logger to collect data, or use software to process data.

You will be assessed by your teacher against the **Common Practical Assessment Criteria (CPAC)**. CPAC defines the minimum standard required for the achievement of a pass.

You will need to keep an appropriate record of your practical work, including your assessed practical activities. This should be done using a practical notebook provided by your teacher. Make sure all your written work in it is clear and well set out — it will form part of your teacher's assessment of CPAC. You will also use it when revising for your final exams.

If you demonstrate the required standard across all the requirements of the CPAC you will receive a 'pass' grade. Your teacher will make a judgement of your practical competence using the CPAC and will need to include comments to justify the decision made. To achieve the pass grade you need to consistently and routinely exhibit the competencies listed in the CPAC before the completion of the A-level course. A major component of the practical endorsement is the ability to independently apply your skills to the practical work. This means that you cannot just rely on your teacher to tell you what to do.

The CPAC are listed below:

1 Follows written procedures.
2 Applies investigative approaches and methods when using instruments and equipment.
3 Safely uses a range of practical equipment and materials.
4 Makes and records observations.
5 Researches, references and reports.

To be awarded a pass you must consistently and routinely meet these criteria. (For further detail on the CPAC, see pages 37–38 of the Edexcel A-level Biology specification.)

This guide deals with each of the core practicals in turn, focusing on the scientific background and the practical techniques involved, including tips to ensure that you meet the relevant CPAC, and conclusions and evaluations from the investigation. The table on the next page lists the core practicals and shows the CPAC and practical techniques that apply to each of them. It also shows the relevant exam-style questions included in the second section of this book.

How the practical element is assessed

Core practicals and linked CPAC, practical techniques and exam-style questions

Core practical		CPAC	Practical techniques	Exam-style question
1	Investigate a factor affecting the initial rate of an enzyme-controlled reaction	1a, 2a, 2b, 3a–3c, 4a, 4b, 5a	1, 2, 3, 6, 8, 12	Question 3: Enzyme-controlled reactions
2	Use of the light microscope, including simple stage and eyepiece micrometers and drawing small numbers of cells from a specialised tissue	1a, 4a	4, 5	–
3	Make a temporary squash preparation of a root tip to show stages of mitosis in the meristem under the light microscope	1a, 2a, 2b, 3a–3c, 4a, 4b	3, 4, 5, 6, 8, 10	Question 10: Cell division
4	Investigate the effect of sucrose concentrations on pollen tube growth	1a, 2a–2d, 4a, 4b, 5b	1, 2, 3, 4, 5, 8	–
5	Investigate the effect of temperature on beetroot membrane permeability	1a, 2a–2d, 4a, 4b, 5b	1, 2, 3, 5, 8	Question 2: Plasma membrane permeability
6	Determine the water potential of plant cells	1a, 2a, 2b, 4a, 4b	1, 2, 3, 5, 8	Question 1: Plant cell water potentials
7	Dissect an insect to show the structure of the gas exchange system	1a, 2a, 2b, 3a–3c, 4a	4, 5, 10	–
8	Investigate factors affecting water uptake by plant shoots using a potometer	1a, 2a–2d, 4a, 4b, 5b	1, 2, 3, 8	Question 8: Transpiration
9	Investigate factors affecting the rate of respiration using a respirometer	1a, 2a, 2b, 4a, 4b	1, 2, 3, 8, 12	Question 6: Respiration
10	Investigate the effects of different wavelengths of light on the rate of photosynthesis	1a, 2a–2c, 4a, 4b, 5b	1, 3, 8	Question 4: Photosynthesis
11	Investigate the presence of different chloroplast pigments using chromatography	1a, 2a, 2b, 3a, 3b, 4a, 4b, 5b	1, 3, 8, 9	Question 5: Photosynthetic pigments
12	Investigate the rate of growth of bacteria in liquid culture	1a, 2a, 2b, 3a, 3b, 4a, 4b, 5b	1, 3, 9, 10	Question 9: Counting bacteria
13	Isolate individual species from a mixed culture of bacteria using streak plating	1a, 2a, 2b, 3a, 3b	3, 9	–
14	Investigate the effect of gibberellin on the production of amylase in germinating cereals using a starch agar assay	1a, 2a–2d, 4a, 4b	1, 3, 6, 8, 9	Question 7: Investigating the effect of gibberellin
15	Investigate the effect of different sampling methods on estimates of the size of a population	1a, 2a–2d, 4a, 4b, 5a	1, 5, 11	–
16	Investigate the effect of one abiotic factor on the distribution or morphology of one species	1a, 2a–2d, 3a–3c, 4a, 4b, 5a, 5b (dependent on investigation)	5, 8, 11, 12	–

Skills Guidance

Practical advice

Following instructions

When carrying out practical work you usually follow a set of written instructions. Before you start the practical work read them through to the end. Annotate the instructions to help you understand them. Then check that you have all the apparatus and materials listed. When you are ready to start read the first instruction carefully and carry it out. Put a tick by each instruction when you have completed it. Proceed carefully through the rest of the instructions, double-checking that you are sticking to the instructions. This is important not only to ensure the collection of accurate data but also that the practical activity is safe. Following instructions shows competency in CPAC 1a.

Safety

Carrying out practical work safely is absolutely essential. If you work unsafely you are putting other people in the class at risk, as well as yourself. Safe working demonstrates competency of CPAC 3a and 3b. Below is some general safety advice, but if you are unsure if something is safe ask your teacher.

- Keep your work area well organised and tidy.
- Use one area for wet work and keep another area dry for writing in your notes/lab book. Do your practical work over the bench, not over your papers.
- Wear protective goggles/spectacles.
- Make sure that you are familiar with hazard warning symbols and know how to respond to them.
- Take special care when using knives, scalpels, glassware, chemicals, Bunsen flames, hot water etc.
- Inform the teacher or technician immediately if you have an accident.
- Clear up any spillages immediately.

Table 1 shows examples of poor laboratory practice that teachers will use as reasons for not awarding you the relevant competency. It also shows equivalent good practice.

Practical tip

Your teacher will assess your approach to working methodically and safely while you carry out your practical work.

Table 1 Examples of poor and good laboratory practice

Poor practice	Good practice
Using incorrect apparatus or equipment without realising*, for example using a $10\,cm^3$ syringe instead of a $1\,cm^3$ to dispense $0.5\,cm^3$ of a liquid	Choosing the correct apparatus/equipment/ chemicals from those provided so that the correct results are obtained
Using the same syringe to dispense different liquids without realising* so that contamination occurs	Using separate syringes when necessary and keeping syringes separate once used (for correct re-use) *or* washing out a syringe between using to dispense different solutions
Cutting slices or cubes or sections carelessly so that sections are of uneven thickness or cubes are of unequal size	Using cutting equipment (e.g. scalpels and knives) and measuring equipment (e.g. rulers) with care to produce slices and cubes of correct sizes
Allowing fluids to drip off the outside of beakers/tubing/stirring rods/ tissue samples into other solutions so that there is the risk of cross-contamination	Rinsing and drying equipment when necessary, clearing off spills on the outside of beakers; keeping different items in clearly defined areas on the bench
Haphazard use of the stop clock/ bench timer so that incorrect times are recorded; samples are not taken at correct time intervals	Checking how to use the stop clock/bench timer before starting; careful checking of times; re-setting back to zero when required
Filtering suspensions through a filter funnel where the filter paper has not been folded correctly/has a tear/has not been fitted into the funnel correctly	Folding a piece of filter paper and then opening it out into the filter funnel; filtering suspensions so that a clear solution, the filtrate, runs through and the precipitate/ larger insoluble particles remain on the filter paper
* Realising a mistake and asking for fresh syringes is considered good practice.	

Practical tip

Syringes should be washed out with water and then with a small volume of the liquid you are going to dispense.

Planning investigations

In the exam papers you may be asked to plan some aspect of an investigation. Obviously you are not going to carry it out, but you should write your answer as if you are. Also you will probably plan and carry out a complete investigation as part of your practical work so you can be assessed on your skills at researching and writing a scientific report.

Throughout your course and in the exams you will be tested on the skills involved in planning, such as identifying variables, stating a hypothesis, writing a method or explaining how to collect results and/or analyse them. Writing full plans during your course will prepare you well for these questions. Here are some steps to follow while planning an investigation.

1 **Identify a question to answer and write a hypothesis**. Read the information provided carefully and look for clues. Write a question that you have to answer by experiment and then write a hypothesis, which is a clear statement about what you think will happen. You must write a **testable hypothesis** — one that you can test by experiment.

2 **Carry out some research**. Use sources of information to read about the problem you are trying to solve. Look for ideas for the strategy you will plan and decide how results could be analysed statistically.

3 **Write a null hypothesis**. If you are going to use a statistical test to analyse your results then you must rewrite your hypothesis as a negative statement, known as a **null hypothesis**. This states the opposite of your hypothesis. Your experiment must test the idea that there will be *no effect*.

4 **Identify the independent, dependent and control variables for the investigation:**
   - The **independent variable** (IV) is the one that you choose to change (or are told to change) in an investigation. In some investigations there are two or more IVs.
   - The **dependent variable** (DV) is the variable that you do not know at the beginning; it is the variable that you set out to observe and/or measure.
   - **Control variables** (CV) are those that are kept constant as they might have an impact on the values of the DV.

5 **Decide on a strategy for your experiment**. This is a brief outline of a method that tests your hypothesis.

6 **Choose apparatus and materials that are appropriate**. You should choose only apparatus that is available in a school or college lab, not sophisticated apparatus such as electron microscopes or DNA sequencing machines that are not. Often it is a good idea to justify your choice of the main items of apparatus.

7 **Expand the strategy into a detailed procedure, using numbered steps**. Avoid using continuous prose because it is easier to follow a series of instructions. Notice that numbered points allow you to include instructions such as 'repeat step 6'. The procedure must describe how results are to be collected.

8 **Explain how results are to be presented**. A good way to do this is to write an outline of a table with columns and rows headed up in full, with units.

9 **Explain how results are to be analysed**. This includes
   - any data processing
   - the type or types of graph (if any)
   - the appropriate statistical test

10 **Plan and carry out a pilot investigation** to trial your ideas. You may have to modify your strategy and/or your detailed plan as a result. If so, record all the details and include them in your report.

Practical tip

If you choose to use a colorimeter, you could say that this gives quantitative results that are easier for others to reproduce than if you use colour standards (charts showing expected colours). You will use a colorimeter in core practical 1.

Evaluating procedures

When carrying out an experimental procedure it is important to consider the way in which the procedure was carried out and the quality of the data collected. You need to ask yourself the question: 'can I have confidence in my data?' If you do not have confidence in the data then you cannot have confidence in the conclusion(s) that you make.

There are quite a few terms with specific meanings that are used when discussing the quality of the procedure and the results obtained:

- **Resolution** is the smallest change in quantity that can be measured with the apparatus that you are using. This refers both to the apparatus used to measure out quantities as part of the procedure and the apparatus used to collect results. It refers to the number of significant figures (or decimal places) in readings.
- **Precision** is a measure of the closeness of agreement between individual results obtained using the same procedure under exactly the same conditions. However, closeness of replicates does not mean that the data are close to the true value.
- **Repeatable results** are replicate results that are in close agreement. You can use mathematical methods to help evaluate the variation in replicate results.
- **Reproducible results** are results that are produced by someone else who follows exactly the same procedure using the same apparatus and materials, but in a different place and at a different time. You can only comment on this in response to a question if you are given results from different people.
- **Accuracy** is a measure of the closeness of agreement between individual results or a set of results and an accepted 'true' value. In biology it is often difficult to know the 'true' value in an investigation. There will usually be a 'true' value, but errors and limitations reduce the chances that results are close to that 'true' value and therefore accurate.

Practical tip

Notice that resolution also applies to microscopy. It refers to the smallest distance that can be detected when using a microscope.

Practical tip

Take special care over using the term accuracy. There are few biology investigations in which you can say what the true value(s) should be.

The first thing to do when evaluating is to consider the procedure that you followed. Is it possible that there were any **measurement errors** in the method? There are two types of error:

- **Systematic errors** are always the same throughout the investigation. A common type of systematic error is when the measuring device gives readings that are incorrect by a certain value. It could be that one of the controlled variables is always incorrect by the same quantity. If there are small systematic errors (that are always the same) then the data may be precise, but not accurate. The effect of these errors is to overestimate or underestimate the true values of the dependent variable.
- **Random errors** occur when you do not carry out the procedure in exactly the same way each time. You may also read the apparatus in a slightly different way each time you take a reading. These errors affect some of the results, but not all of them. They do not always affect the results in the same way. Random errors could be the result of the variation in biological material.

Practical tip

The words in bold type are important key terms that you must understand and use. Write yourself a list with their definitions.

Do not think of errors as mistakes. Even in a perfectly performed investigation there will be errors. Systematic errors may not be easy to identify, but you should always check the accuracy of any measuring instruments, such as balances, colorimeters and pH meters. Random errors should show up in the data, making the data less reliable. However, random errors may affect one value of your independent variable, but not all of them (Figure 1).

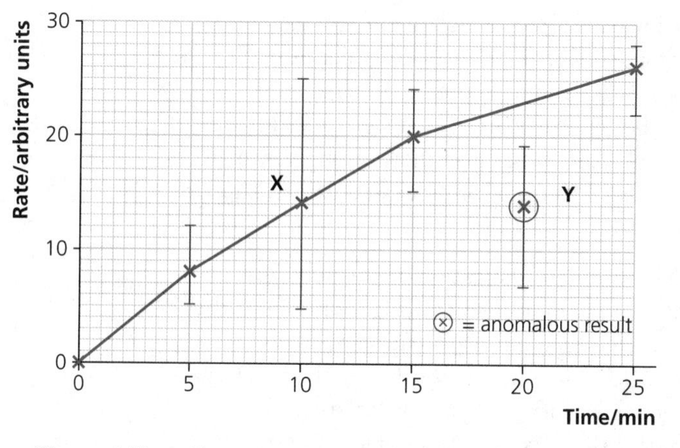

Figure 1 The effect of random errors on processed data

The results plotted are mean values with range bars. We can see that the result labelled X has a much wider range than the others. This suggests that there may have been one, or more, random errors in the readings for 10 minutes. The result labelled Y does not fit the overall trend of the other results. This may be because there were random errors when collecting all the data for 20 minutes.

Anomalous results are results that do not fit the trend. They are sometimes known as outliers. An anomalous result can be:

- a replicate result that differs significantly from the others
- a result (which may or may not be a mean) for one value of the independent variable that does not fit the overall trend and is not included in the curve of best fit

You may be asked to suggest likely explanations for anomalous results. The anomaly could be the first result taken before the experimenter was confident in the procedure.

Validity refers to individual measurements and to the whole procedure. If you have a **valid result**, then you know that you measured what you set out to measure. If you have a **valid investigation** then you have measured what you intended to measure and you can be confident that the effect of changing the independent variable leads to changes in the dependent variable.

When you make a conclusion about an investigation then you can make a judgement about the extent to which the evidence collected supports that conclusion. In so doing you are expressing **confidence** in your conclusion. If asked to comment on the confidence in a conclusion then you should consider the following:

- the limitations in the procedure
- any uncontrolled variables
- the effects of errors (systematic and random) on the results
- the repeatability of the results
- the precision of the data collected
- the accuracy of the results

Give some positive aspects of the investigation first, followed by some criticisms. You should refer to specific aspects of the procedure and results, rather than using vague comments such as 'my conclusion is valid because my results are precise, reliable and accurate' — this is meaningless without supporting information. For example, you can say that your results are precise because you measured to two decimal places and that they are reliable because the replicates are close together and there are no anomalous data. Your results may be accurate because they all agree with an expected trend. Always quote some examples of your raw or processed data in support of your arguments.

You should consider the variables involved in the investigation. Controlled variables are all the other variables that you were not investigating. Sometimes the instructions will tell you to keep these constant. In an investigation of enzyme-catalysed reactions, the temperature may be kept constant by placing the reaction mixtures in a thermostatically controlled water bath, for example at 20°C. The pH may also be controlled by using a buffer solution of, for example, pH 7.0. If variables are not controlled then they may influence the results; they are called **confounding variables** or **uncontrolled variables**. Sometimes, as in field studies, you may be aware of such variables and then 'take them into account' when analysing and interpreting results.

Controlled variables should not be confused with a **control experiment**. It is important to know that your results are valid, that they show what you think you are investigating.

Repeatability and accuracy

In A-level practical tasks it is usual to carry out three repeats or replicates for each value of the independent variable if time and materials permit. These should be done separately from each other using exactly the same experimental procedures.

A balance used for weighing sections of potato tuber tissue might measure to the nearest 0.1 g. Some balances are more sensitive and weigh to the nearest 0.01 g. If measuring change in length you will use a ruler that measures to the nearest millimetre. These are measures of **resolution** in results taking. Recording to the nearest 0.1 g gives a higher resolution than measuring to the nearest 1 g. Similarly, measuring to the nearest millimetre gives a higher resolution than measuring to the nearest centimetre. So 10.6 g has a higher resolution than 11 g and a 13 mm measurement has a higher resolution than a 1 cm measurement. In this context resolution means the number of significant figures or decimal places to which values are expressed.

Stop clocks and bench timers can often measure to a hundredth of a second (0.01 s). It is highly unlikely that you could time a colour change or other event to this degree of resolution so it is better to express your results to the nearest second (or even to the nearest 15 seconds or 30 seconds).

Resolution does not only involve results taking. It also involves the resolution of the apparatus that you use for preparing materials for practical tasks. For example, you may use a syringe for measuring out volumes. It is not easy to measure exactly with a syringe especially with a coloured solution. It is possible to improve this by using a graduated pipette or a burette.

Practical tip

Calculating a **running mean** is a good way to check that you have enough replicate results. Calculate the mean after you have collected each replicate and continue doing this until it remains near constant.

Uncertainty in measuring

Uncertainty is half the smallest graduation on the apparatus — for example, if the smallest division on a syringe is 1.0 cm^3 then the uncertainty would be ± 0.5 cm^3. So if you start measuring at 0 the uncertainty applies where you take the measurement — say at 6.3 cm^3. The result is expressed as 6.3 ± 0.5 cm^3. But if you have to start at a measurement other than 0 (for example when taking readings from a burette) the uncertainty applies at both ends, so it is multiplied by two as there is an error at each end, for example 7.5 ± 1.0 cm^3. The same applies to measuring a quantity in a syringe by sucking up from empty — the error would be half the minimum measurement. But when you take two readings from the syringe (say delivering 2.0 cm^3 by moving the plunger from 6.5 cm^3 to 4.5 cm^3) then the uncertainty is multiplied by two.

It is possible to calculate the **percentage error** for the apparatus you used for measuring your results. Imagine that you have collected a gas and measured the volume with a gas syringe that has graduations every 1 cm^3. If you have started from zero and measured 5 cm^3 of gas with your syringe, you can be certain that you have more than 4.5 cm^3 but less than 5.5 cm^3. Your error is ± 0.5 cm^3 in 5 cm^3. This makes the percentage error:

$$\text{percentage error} = \frac{0.5}{5.0} \times 100 = 10\%$$

If the volume of gas collected was 10 cm^3, then the percentage error would be 5%.

Accurate data

It can be challenging to determine how accurate the data you collect in A-level practicals are. In biological investigations the true value is not always known. In some cases results can be checked with sources of data. For example, tidal volume readings should be about 500 cm^3. The water potential of the blood should be equivalent to 0.9% sodium chloride solution (−3.86 MPa). But the water potential of plant tissues varies considerably and there is no specific value against which results can be checked.

This is why it is important to evaluate the procedure followed in an investigation and evaluate the results obtained. Results are often considered to be anomalous and you need to think about how these might have been obtained. However, what might seem to be anomalies or outliers are not necessarily so.

Recording observations and results in tables

In most of your practical work you will need to record results and observations. In almost all cases, results and observations will be recorded in tables.

Before you start to draw a table, decide what you wish to record. Decide on how many columns and how many rows you will need, and make a rough table in pencil. Make sure that you have read all the practical instructions before you draw the table outline. Follow these rules:

- Use plenty of space — do not make the table too small.
- Leave some space to the right of the table in case you suddenly decide you need to add more columns.
- Make the table ready to take observations and/or readings so that you can write them directly into the table rather than writing them on another page and then copying them into the table. Tables should show all the raw data you collect.

Practical tip

There are plenty of examples in this guide to help you become proficient in drawing and completing tables of results.

- Use a pencil and ruler to draw lines between the columns and between the rows. Rule lines around the whole table.
- Write a brief, but informative, heading for each column.
- The headings of the columns and rows that record measurements must include the relevant units. Write pure numbers in the body of the table without any units.
- When two or more columns are used to present data, the first column should be the independent variable; the second and subsequent columns should contain the dependent variables.
- If you have collected replicate results then they should all be shown.
- Entries in the body of the table should be brief — they should be single words, short descriptive phrases, numbers, ticks or crosses.
- Data should be ordered so that trends and patterns can be seen — it is best to arrange the values of the independent variable in ascending order.
- Tables should be given informative headings.
- Units are separated from the description of the variable by a forward slash (/) or by putting the units into brackets. The slash should *not* be used to mean 'per' in compound units. For example, if you include concentrations as in Table 4 (page 30) do *not* write mol per dm^3 as mol/dm^3. It should be written as mol dm^{-3}.

Practical tip

Sometimes it may be necessary to draw the table in landscape, especially if you are collecting a lot of raw data.

Core practicals

Core practical 1

Investigate a factor affecting the initial rate of an enzyme-controlled reaction

CPAC 1a, 2a, 2b, 3a–3c, 4a, 4b, 5a

Background

The rate of an enzyme-controlled reaction can be calculated by measuring the change in the concentration of the product or the concentration of the substrate. The graphs in Figure 2 show how the concentration of (a) product and (b) substrate change over time.

The rate of a reaction can be determined by measuring the rate of change of substrate or product over time.

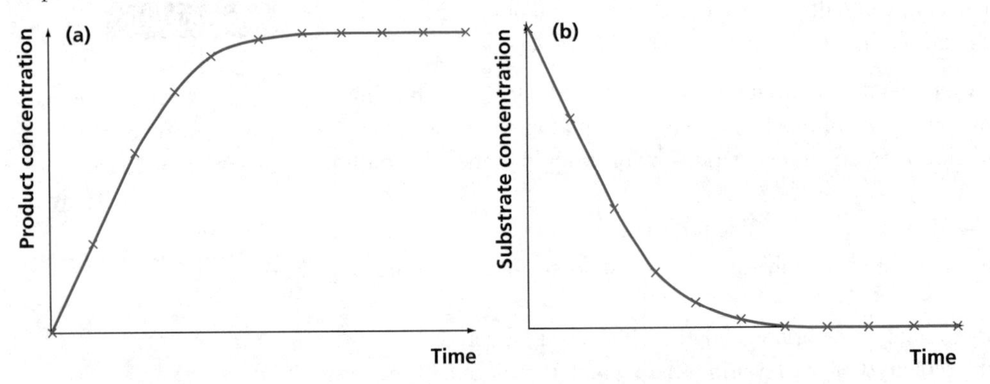

Figure 2 Change in concentration of product and substrate over time

Practical notes

In this required practical the independent variable will be a factor that affects the rate of an enzyme-controlled reaction. This could be temperature, pH, substrate concentration or enzyme concentration. In this example we will be using the concentration of the enzyme trypsin. Trypsin is used to break down the casein protein in milk. As this protein gives the milk its white colour we can measure the rate of reaction by observing how long it takes for the milk to become clear.

Observing the milk going clear is not particularly accurate because observing a colour change is subjective. This means that it is difficult to produce reproducible results because different people will judge the end point differently. It is also difficult for you to obtain repeatable results as you may find it challenging to always judge the end point the same every time. In order to remove the subjective element we can use a device called a colorimeter (Figure 3). A colorimeter measures the absorbance or transmission of light through a solution.

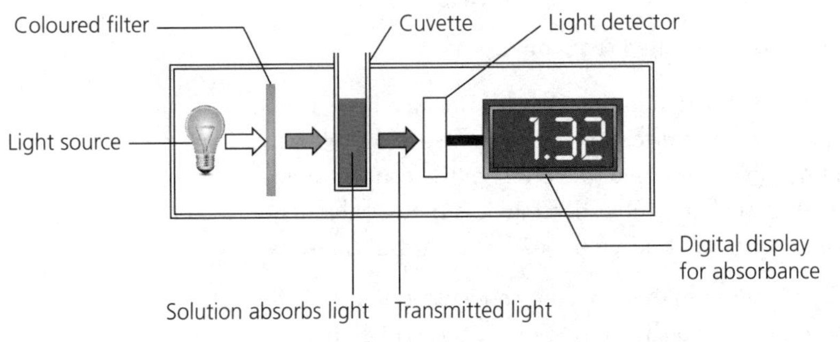

Figure 3 A colorimeter

In this investigation the colorimeter is used to measure the absorbance of light. The absorbance will decrease over time as the trypsin catalyses the breakdown of the protein. Once the absorbance has reached a relatively constant value this indicates the end point. The dependent variable in this investigation is therefore the absorbance of light.

You will be provided with a 1% stock solution of trypsin that you will need to dilute. To do this, add distilled water. Each solution needs to have a volume of $10\,cm^3$.

You may be able to quickly work out the volumes of distilled water and trypsin solution required. If not multiply the concentration by the total volume; this gives the volume of trypsin required. Now add enough distilled water to make the solution up to $10\,cm^3$. Table 2 shows an example of how to do this.

Table 2 Dilutions required to give different concentrations of trypsin solution

Percentage concentration	Volume of 1% trypsin solution/cm^3	Volume of distilled water/cm^3
1.0	10	0
0.8	8	2
0.6	6	4
0.4	4	6
0.2	2	8

As this experiment involves a fixed volume of substrate (protein in the milk), the substrate will quickly become the limiting factor (the factor that limits the rate of reaction). The rate of reaction over time is shown in Figure 4. As you can see, the rate of reaction starts at a high level before decreasing as the reaction progresses and the substrate is converted to product. Once the rate of reaction begins to decrease, the substrate has become the limiting factor.

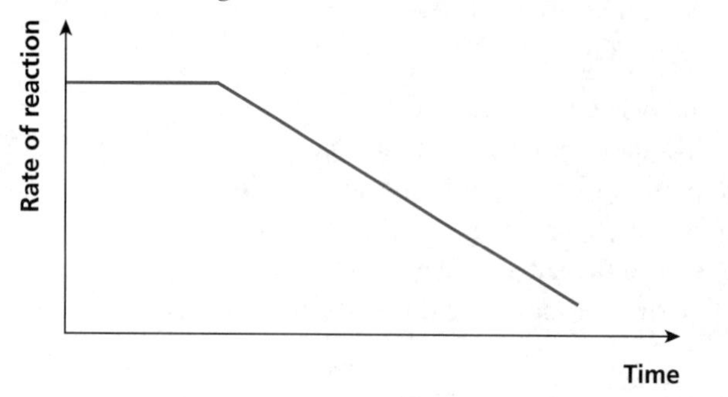

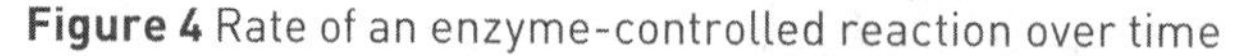

Figure 4 Rate of an enzyme-controlled reaction over time

It is important to measure the initial rate of reaction in this investigation, as this is the point where it is the enzyme concentration that is affecting the rate of reaction. Once the rate of reaction begins to fall it is not possible to separate the effect of the concentration of substrate from the effect of concentration of enzyme. Therefore it is the initial rate of reactions that can be used to compare the effect of different enzyme concentrations.

The initial rate of reaction should be higher at higher enzyme concentrations. This is because there are more available active sites and therefore more successful collisions will occur between active sites and substrates, reducing the concentration of substrate and so decreasing the absorbance.

To calculate the initial rate of reaction, draw a graph of absorbance over time for each concentration of trypsin solution (you should be able to fit all lines on one graph). You can then use a tangent to calculate the initial rate of reaction (this should be the steepest point on the curve). See Figure 5.

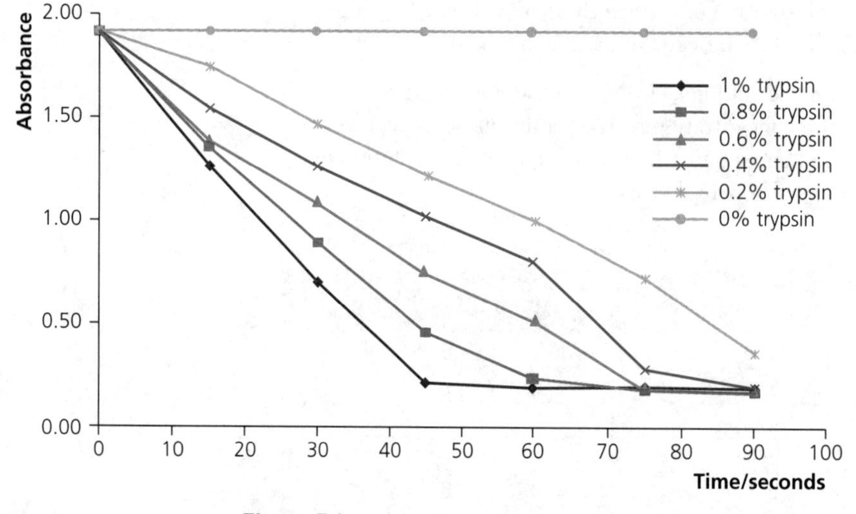

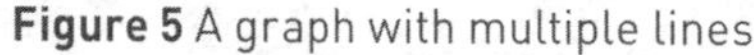

Figure 5 A graph with multiple lines

Practical tip

When drawing line graphs with multiple lines, use different symbols or different types of lines for each variable (as shown in Figure 5) to ensure that your graph is clear and readable.

Focus on maths skills

Tangents

A **tangent** is a straight line that touches a graph at a single point. The gradient of this line will give the rate of change (rate of reaction) at that point on the line.

To calculate the gradient the tangent is used as the hypotenuse of a right-angled triangle. Dividing the height of the triangle by its length gives the gradient of the hypotenuse and therefore the gradient of the tangent.

Once you have calculated the initial rate of reaction for each of the concentrations, a graph of initial rate of reaction over trypsin concentration can be drawn.

Core practical 2

Use of the light microscope, simple stage and eyepiece micrometers, and drawing cells from a specialised tissue

CPAC 1a, 4a

Background

In this practical you will be observing and drawing sections of biological samples using the low-power and high-power lenses of a microscope. This practical uses plant stem as an example tissue.

Practical notes

Students can often be quite intimidated by microscopy — don't be! Although it may seem difficult at first, with care and practice it is possible to become confident in using a microscope.

There are usually three or four objective lenses on the microscopes you will be using, with magnifications of ×4, ×10, ×40 and ×100. The image formed by these lenses is further magnified by the eyepiece lens, which is usually ×10. This therefore gives overall magnifications of ×40, ×100, ×400 and ×1000. Generally ×40 and ×100 are considered low power while ×400 and ×1000 are high power.

Microscope lenses should be clean before you use them. The eyepiece lens and the objective lenses should be cleaned with special lens tissues. Be careful how you hold and use the microscope so that you do not leave greasy finger marks on the lenses.

Here is some advice on using a light microscope to view prepared slides:

- If there is a condenser on the microscope, focus this for each objective lens. Do this by placing a slide on the stage and a pencil point on the light source, and adjust the condenser until both are in focus.
- Look at the slide with your naked eye first before putting it on the microscope. This will help you position the slide on the stage of the microscope.
- Always use the low-power objective lens (×4) first.

Practical tip

The ×100 lens is an oil-immersion lens so can only be used when a drop of oil is placed onto the slide and the lens carefully lowered into it.

- Once positioned on the stage, move the slide so that it is centred beneath the lens and use the focusing wheel to bring the object into focus. If your microscope has translation control knobs use them to move the stage rather than pushing the stage/slide around.
- Move the slide around to see everything that is beneath the coverslip and then move the next objective lens (×10) into position to view something of interest.
- Focus the microscope by turning the wheel so that the lens moves upwards or the stage moves downwards according to the type of microscope. Search the slide by moving it on the stage.
- When you find something that you want to look at under high power, move the ×40 objective lens into place. To focus, move your head so that your eyes are on the same level as the stage and look carefully at the gap between the objective lens and the coverslip. Carefully reduce the gap by lowering the lens or raising the stage so the lens is as close as possible without touching the coverslip. Look through the eyepiece lens and focus by raising the lens or lowering the stage, being careful not to touch the slide with the lens.

Measurements of cells and cell structures are made with an eyepiece graticule. This is a graduated scale printed onto a piece of plastic that is inserted into the eyepiece. Before the scale is used it must be calibrated. This is because the actual distance between the divisions on the scale depends on the magnification. You can see this for yourself by looking down the microscope and selecting different objective lenses. As you increase the magnification of your microscope each division represents a smaller length. A stage micrometer is a special microscope slide that has a very small scale printed on it, usually with divisions that are 0.1 mm apart and 0.01 mm apart.

Here are some instructions on calibrating the eyepiece:

- Put the stage micrometer on the stage of the microscope and align the zero on the eyepiece graticule with zero on the stage micrometer.
- Note the number of eyepiece divisions that matches up with 10 stage micrometer divisions. You may have to adjust this depending on the magnification and if an eyepiece unit does not exactly match up with a stage micrometer division.
- Calculate the actual distance between the divisions on the eyepiece graticule. For example, at ×40 magnification 40 eyepiece divisions = 10 stage micrometer divisions.
- As each micrometer division = 100 µm (this is standard for most micrometers but you may need to adjust it depending on the micrometer you are using):

$$1 \text{ eyepiece division} = \frac{\text{number of stage micrometer divisions}}{\text{number of eyepiece divisions}} \times 100$$

$$1 \text{ eyepiece division} = \frac{10}{40} \times 100 = 25\,\mu\text{m}$$

- This answer can also be expressed in mm in standard form. 1 mm = 1000 µm, so:

$$25\,\mu\text{m} = \frac{25}{1000} = 0.025 = 2.5 \times 10^{-2}\,\text{mm}$$

When measuring the width of an object such as a cell, position the eyepiece graticule over the object and count the number of eyepiece graticule divisions. You can record the length of the object you are measuring in number of eyepiece divisions or eyepiece units (EPU). At high power (×400) each small division on the eyepiece scale (1 EPU) should be equivalent to around 2.5 μm.

Calculate the actual length using this formula:

actual length = number of eyepiece units × calibration

For a cell of length 25 eyepiece units and a calibration of one eyepiece unit equalling 2.5 μm:

actual length of cell = 25 × 2.5

= 63 μm (2 s.f.)

Keep a record of the calibrations at each magnification, for example ×40, ×100 and ×400. This means that you will not have to use the stage micrometer each time you take measurements with the eyepiece graticule.

Magnification

To calculate the magnification of a drawing or an electron micrograph use the following equation:

$$\text{magnification} = \frac{\text{image size}}{\text{actual size}}$$

Remember to convert your measurements so that they are both in the same units.

You can rearrange this equation to allow you to calculate an actual size if you know the image size and the magnification.

Focus on maths skills

Example 1

A drawing of a cortical cell was 25 mm wide. The actual cell was 300 μm wide. What was the magnification of the drawing?

Step 1: convert both lengths to the same unit:

$25\text{mm} = 25\,000\,\mu\text{m} = 2.5 \times 10^4\,\mu\text{m}$

Step 2: substitute the values into the magnification formula:

$$\text{magnification} = \frac{\text{image size}}{\text{object size}}$$

$$= \frac{2.5 \times 10^4}{300}$$

= 83 (2 s.f.)

Hence the magnification is ×83.

Example 2

In a diagram of a prokaryotic cell, its width was 2.6 cm. The magnification was given as ×520. What was the actual width of the cell?

In this question you need to find the object size given the magnification and the image size.

Step 1: rearrange the magnification formula to make 'object size' the subject of the equation:

$$\text{magnification} = \frac{\text{image size}}{\text{object size}}$$

$$\text{magnification} \times \text{object size} = \text{image size}$$

$$\text{object size} = \frac{\text{image size}}{\text{magnification}}$$

Step 2: substitute the given numbers into the rearranged formula:

$$\text{object size} = \frac{\text{image size}}{\text{magnification}}$$

$$= \frac{2.6}{520}$$

$$= 0.005\,\text{cm} = 50\,\mu\text{m}$$

Practical tip

Always measure your drawing in millimetres (mm) and express your answers in standard form or convert your answers to micrometres (μm).

In this practical you may use a stain to more easily view the cells and tissues of the stem. Toluidine Blue O is a metachromatic stain, which means that it reacts differently with different chemical components of cells, producing a variety of colours. Toluidine Blue O stains lignin and tannins green to blue, pectins pinkish-purple, and nucleic acids purplish or greenish blue. This stain can therefore be used to show the presence of these components in areas of the cell, allowing us to draw conclusions on both structure and function. Other stains that can be used include iodine and Janus Green B.

As part of this practical you will need to make drawings to show the arrangement of tissues within the stem and also of individual cells. The most important thing to remember when carrying out biological drawing is that you do not need to be brilliant at art to be able to do it. You will need to do a low-power plan drawing of the tissues in the stem. The purpose of this is to show the arrangement and thickness of different tissues, but not to show the cells that make up these tissues. For these plan drawings you should use either the ×4 or ×10 objective lens to view slides.

Practical tip

When you write handwritten scientific names they should be underlined. In print they are in italics.

To show the arrangement of cells at the junction of two tissue types you will produce a high-power plan drawing. For this you will need to use a high-power lens (×40).

For a low-power drawing, follow these rules:

- Make the drawing fill at least half the space provided; leave space around the drawing for labels and annotations (see advice about this below).
- Use a sharp pencil, such as an HB (never use a pen), a good-quality sharpener and a clean eraser.
- Use thin, single, unbroken (clear and continuous) lines. It can be a good idea to pull rather than push the pencil as you draw.
- Show the outlines of the tissues without drawing any cells.
- Make the proportions of tissues in the diagram the same as in the section.
- Do not use any shading or colouring.
- Give the drawing a suitable title and always include the scientific name of the organism concerned if known.

Practical tip

Low-power drawings are often called tissue maps. They show the distribution and arrangement of tissues in organs. You can practise this skill by drawing suitable images taken from the internet.

When drawing a high-power diagram, observe the same rules as for the low-power plan (except that here you will be drawing individual cells), but with the following additions:

- Draw only a few cells that are representative of the tissue that you are studying. All the cells in the parenchyma are the same, so draw no more than three adjoining cells in as much detail as you can. Do not draw the cells in isolation; instead show how they are attached to each other.
- Make sure that the proportions of cells in the drawing are the same as in the section you are drawing, i.e. draw them with the same width:length ratio.
- Use two lines to indicate the cell walls of plant cells.
- Show any details of the contents of cells — draw what you see, not what you know should be present.
- You will usually be expected to add labels and annotations (notes) to your drawings. Use a pencil and a ruler to draw straight lines from the drawing to your labels and notes. Write labels and notes in pencil, never in pen, in case you make a mistake.
- Your high-power drawing could be annotated to show how cells in the stem are adapted to their function.

Figure 6 shows high-power detail of cells of xylem vessels and phloem sieve tubes.

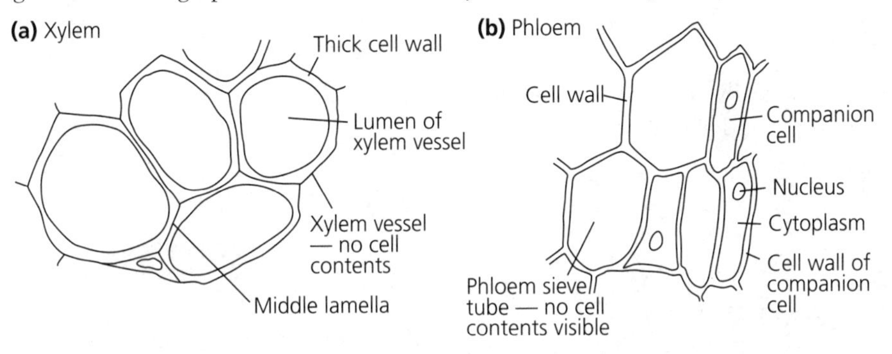

Figure 6 Annotated high-power drawing of a plant stem

Core practical 3

Make a temporary squash preparation of a root tip to show stages of mitosis in the meristem under the light microscope

CPAC 1a, 2a, 2b, 3a–3c, 4a, 4b

Background

Cell division is the process whereby a cell splits to form daughter cells. Before cell division occurs nuclear division occurs. There are two types of nuclear division: mitosis and meiosis. In this investigation you will be studying cells undergoing mitosis.

Chromosomes

One of the most important aspects of cell division is the behaviour of the **chromosomes**. A chromosome consists of DNA and protein. A gene is a specific section of the DNA in chromosomes that codes for one polypeptide. Each chromosome is made up of two identical chromatids joined by a centromere. Chromosomes form **homologous** pairs. Each chromosome in a pair contains the same genes but in different forms. Figure 7 shows a homologous pair of chromosomes.

Exam tip

Many similar sounding words are used in cell division (e.g. chromosome, chromatid, centromere, centriole). Make sure you are clear about what each one means.

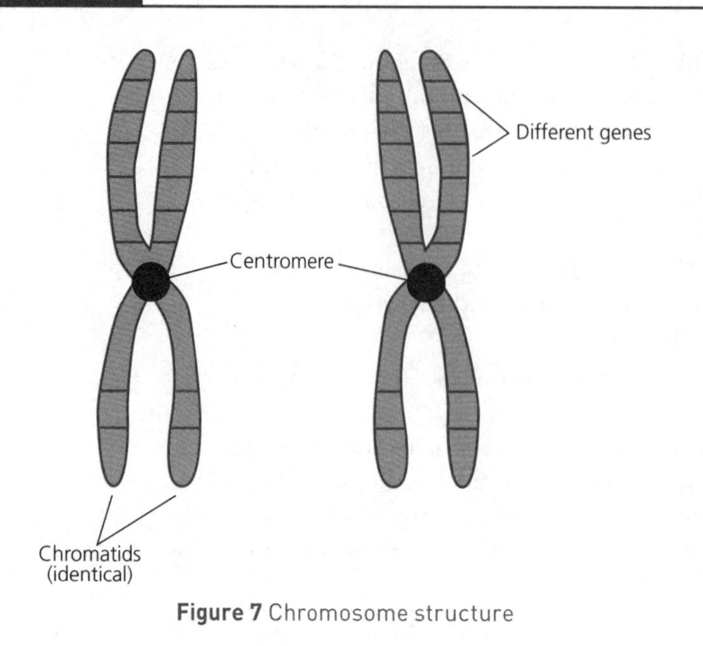

Figure 7 Chromosome structure

Mitosis and the cell cycle

Mitosis produces two cells that are genetically identical to each other and to the parent cell. This provides genetic stability.

The cell cycle is shown in Figure 8.

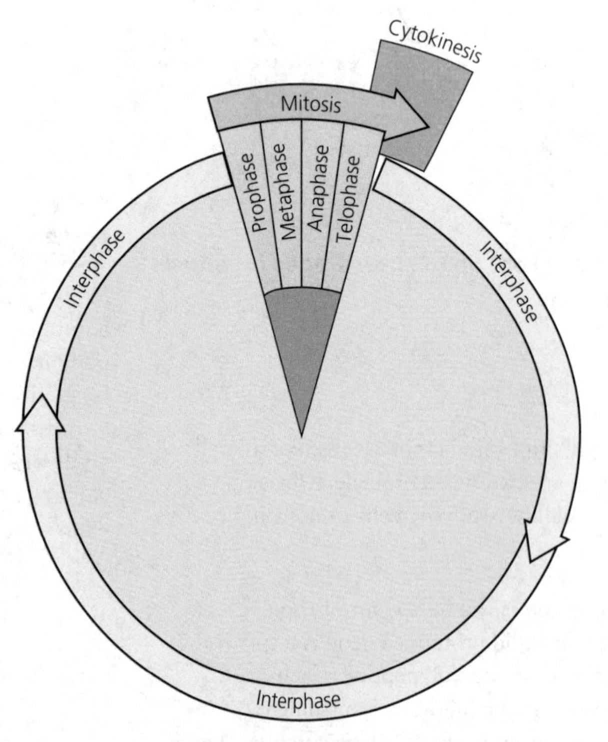

Figure 8 The cell cycle

Exam tip

Remembering the correct order of the stages of mitosis is fundamental to understanding and explaining the process, but students often mix them up. A mnemonic can be a good way of remembering the order, for example Paul Makes Awful Toast or P-MAT (Prophase, Metaphase, Anaphase, Telophase).

When a cell is not undergoing cell division it is in **interphase**. During interphase the following processes occur:

- DNA replicates
- Protein synthesis occurs
- ATP is synthesised
- Organelles are produced

At the end of interphase mitosis occurs. After mitosis the cell divides by **cytokinesis** to form two genetically identical daughter cells.

Mitosis can be divided into four stages — **prophase**, **metaphase**, **anaphase** and **telophase** (Figure 9). It is important to realise that mitosis is a continuous process and we divide it up into separate stages to aid our understanding of it.

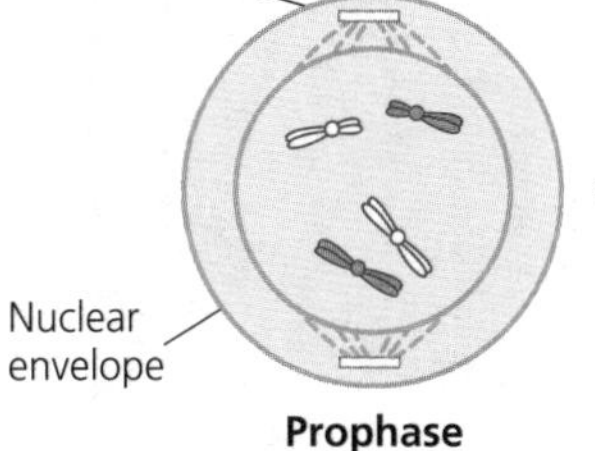
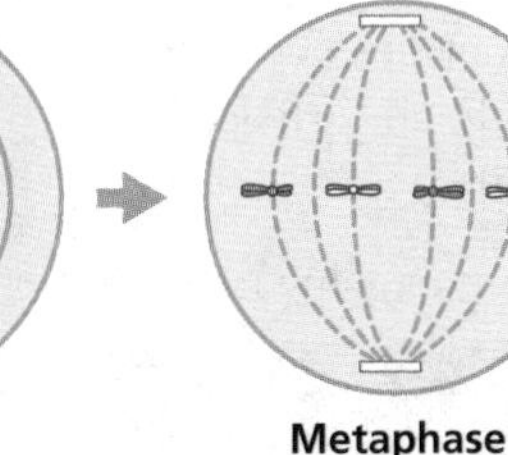

Figure 9 The stages of mitosis

Prophase

This is the longest stage of mitosis. This means that in a sample of cells undergoing mitosis a large number of cells will be in prophase. The following events occur during this stage:

- The chromatin, which is free in the nucleus, condenses to form the chromosomes (two chromatids joined at the centromere).
- Centrioles move to the poles (either end) of the cell and begin to form the **spindle fibres**. The spindle is a web of protein microtubules that extend across the cell.
- The nuclear membrane breaks down.

Metaphase

During metaphase the chromosomes line up at the equator of the cell (in the middle) and the spindle fibres attach to the chromosomes at the centromere.

Anaphase

During anaphase the spindle fibres contract, the centromere separates and the chromatids (now called sister chromosomes) are pulled to the opposite poles of the cell. Anaphase is the shortest stage of mitosis. This means that if you observe a sample of cells undergoing mitosis you will see very few cells in the anaphase stage.

Telophase

During telophase the chromosomes unwind back to chromatin. The nuclear envelope reforms and the spindle breaks down.

Exam tip

You must be able to describe and draw all the stages of mitosis. If you are asked to draw one or more of the stages you will often be given a number of chromosomes to include, so make sure you stick to this number.

Exam tip

When drawing a cell undergoing anaphase it is important you draw the chromosomes as 'V' shapes with the point of the V (and the point of spindle attachment) facing the pole of the cell it is being pulled towards.

Cytokinesis

At the end of telophase the cell divides by **cytokinesis** to form two daughter cells. In animal cells this is done by a 'pinching in' of the plasma membrane. In plant cells a cell plate forms between the dividing cells. This then forms the new cell wall of the two cells.

Functions of mitosis

Mitosis is important for the growth of organisms, the repair of damaged tissues and the replacement of dead cells. In some organisms it is also used for **asexual** reproduction.

Practical notes

In this investigation you will be sampling cells that are undergoing mitosis. While all the processes detailed in the background section will be occurring, your focus will be on the behaviour of the chromosomes and these will be what is visible within the cells.

You will be using the roots of garlic bulbs as a source of cells. The root tips of plants contain regions known as the meristem, where growth is occurring. Each garlic cell only has eight chromosomes, making their behaviour easier to identify.

Figure 10 shows a section of garlic root tip cells, some of which are undergoing mitosis.

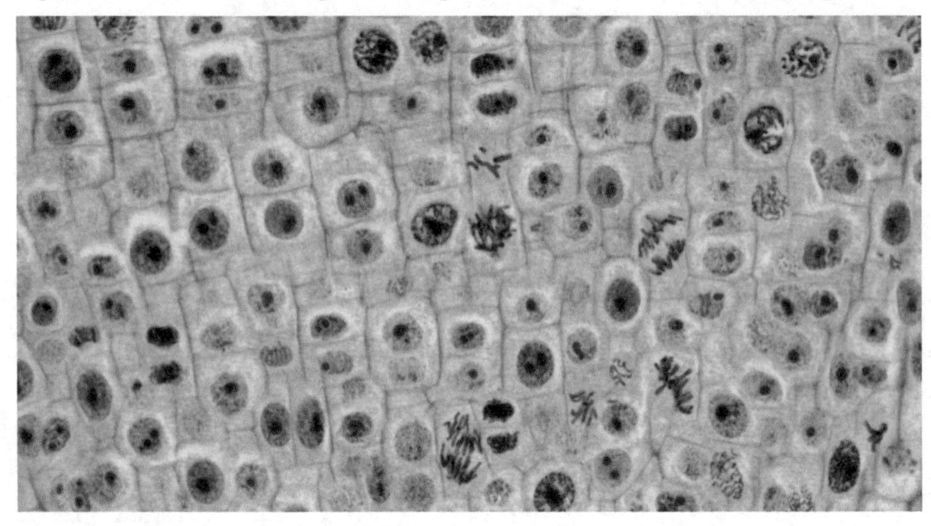

Figure 10 Garlic root tip cells

In order to view the chromosomes properly the cells need to be separated to give a one-cell-thick layer. The plant cells are stuck together by a middle lamella of pectins. You will use hydrochloric acid to hydrolyse the pectins. You will then use acetic orcein stain, which will stain the chromosomes dark red. The stain will also 'fix' the cells. This means that mitosis will stop.

You should identify the stages of the cell cycle by observing the chromosomes and comparing them to Figure 10 and other reference photographs. If no chromosomes are visible and a nucleus is visible this suggests that the cells are in interphase and chromatin has not yet condensed to form chromosomes.

It can be challenging to identify the stage of mitosis — it is important to remember that mitosis is a continuous process and that dividing it into stages is just a way of making it easier for us to understand. This means that many of the cells may show features that are intermediate between two stages. If this occurs you should select the stage that this intermediate stage most appears like. Start by finding the cells on low power and then use high power to observe the position of the chromosomes.

You should make annotated drawings of cells in each of the stages of the cell cycle. The chromosomes are the most important parts of these drawings, so just draw a simple cell outline and do not worry about other structures that are likely not to be visible anyway. Your aim should be to show the relative sizes and positions of the chromosomes as accurately as possible.

You should record the number of cells in each phase, including interphase, in a table. Using these data you can then calculate the mitotic index of your sample and the approximate time spent by the cells in each phase. Some example data are shown in Table 3.

Table 3 Number of root tip cells in each stage of the cell cycle

Phase	Number of cells	Time spent by cells in each phase/%
Interphase	101	76
Prophase	19	14
Metaphase	6	5
Anaphase	2	2
Telophase	5	4

Focus on maths skills

Mitotic index

Calculate the mitotic index using the following equation:

$$\text{mitotic index} = \frac{\text{number of cells in mitosis}}{\text{total number of cells observed}}$$

number of cells in mitosis (those not in interphase) = 19 + 6 + 2 + 5 = 32

total number of cells = 133

Therefore:

$$\text{mitotic index for these data} = \frac{32}{133} = 0.24$$

To calculate the time spent by the cells in each phase divide the number of cells in each phase by the total number of cells visible. We can do this as the number of cells in each phase is a good guide to how long the cells are in that phase, i.e. the longer a phase the more cells should be in it at any one time.

Core practical 4

Investigate the effect of sucrose concentration on pollen tube growth

CPAC 1a, 2a–2d, 4a, 4b, 5b

Background

In this investigation you will be studying the effect of the concentration of sucrose solution on the germination of pollen grains and the resulting growth of a pollen tube.

Pollen grains contain the male gametes of flowering plants and are produced in the anther. The wall of the pollen grain is made up of two layers, the inner intine, which is composed of cellulose, and the outer exine, which is a tough layer that prevents desiccation (drying out). This increases the chance of a pollen grain surviving the movement from the anther to the stigma of another flower.

Once the pollen grains are mature and ready to be released from the anther, the anther dries out and splits open along lines of weakness. This is called dehiscence. The pollen grains are now exposed to the environment and can be picked up by, for example, insects.

When the pollen grain is transferred to the stigma of another flower, the pollen grain then germinates. The exine of the pollen grain ruptures and the tube nucleus controls the growth of a pollen tube. In order for germination to occur stimulus by a sucrose solution is required. The solution is produced by the stigma when a suitable pollen grain lands upon it.

The pollen tube digests its way through the style by secreting hydrolytic enzymes. The growth of the pollen tube is controlled by a tube nucleus and the growth is chemotropic (growing towards chemicals released from the ovary). The male, generative nucleus from the pollen divides by mitosis to produce two haploid gametes, which travel down the pollen tube (Figure 11).

The pollen tube enters the embryo sac through the micropyle (a small hole at the base of the ovule). The male nuclei enter the ovule and fertilisation occurs. One haploid male nucleus fuses with the haploid female nucleus to form a diploid zygote. The other male nucleus fuses with both polar nuclei to form a triploid primary endosperm nucleus. This is known as double fertilisation (Figure 12).

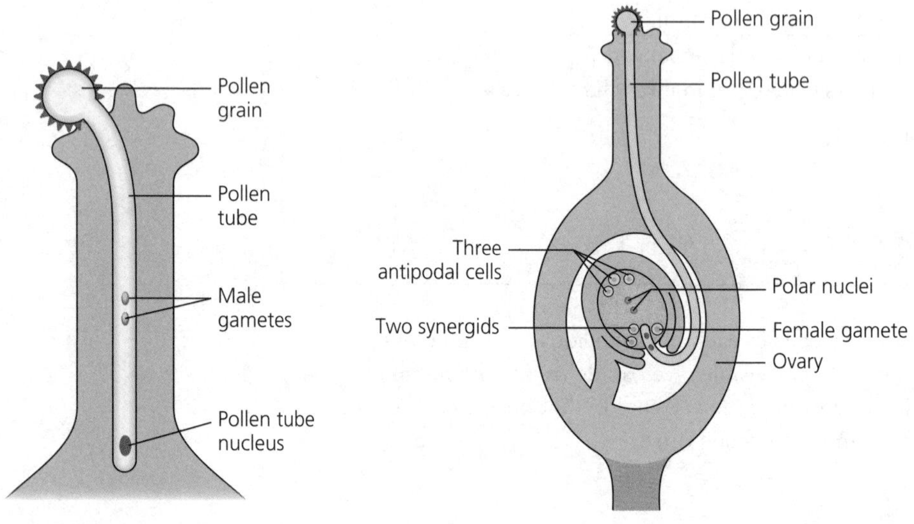

Figure 11 Growth of the pollen tube

Figure 12 Double fertilisation

Practical notes

The independent variable of this investigation is sucrose solution concentration. You should research potential suitable concentrations and then cite the source of your findings. This may involve using primary or secondary sources.

Primary sources are sources of information that are written by people who carried out the research. Almost all of those that you might use are scientific articles published in journals that have been subject to peer review. You might also find pieces of work carried out by students at high school, college or university, which are unlikely to have been peer reviewed, but might be useful sources of information and data. Many articles from scientific journals are freely available on the internet, but many are only available with log-in details from an academic institution such as a university. Your school or college librarians should be able to help you find suitable material.

Secondary sources are books or articles written by people who may have carried out the original research themselves, but more likely did not. These sources could be newspaper reports or articles in science journals such as *New Scientist* or *Biological Sciences Review*. Textbooks and study guides like the ones in this series are secondary sources.

Follow these steps when you carry out your research.

1 Record details of your research, including dates, when you access information and full details of all the sources that you consult, even those that you did not use.
2 Make thorough notes so that you only need to consult each source once.
3 Cite the information you use in your text either as footnotes or as a bibliography.

   A source from an *academic journal* should be cited in the format below:

   Author name(s) (Year), 'Title of paper/article', *Journal*, Issue, Page number(s).

   For example:

   Chebli, Y. and Geitmann, A. (2007), 'Mechanical principles governing pollen tube growth', *Functional Plant Science and Biotechnology*, **1**: 232–245.

   Or from a *book*:

   Author name(s) (Year), *Title of book*, Publisher.

   For example:

   Moore, P. D., Collinson, M. and Webb, J. A. (1994), *Pollen Analysis*, Wiley-Blackwell.

   Or from a *website*:

   Author name, 'Article title', *Website Title*, url, Publisher, (Date published), Date accessed.

   For example:

   Battista, J., 'Pollen tube: growth, function and formation', *Study.com*, www.study.com/academy/lesson/pollen-tube-growth-function-formation.html, Study.com (2013), 12 August 2016.

Once you have determined a suitable range of sucrose concentrations you should prepare small volumes of each of the concentrations of sucrose solutions. You only need to add a few drops of sucrose solution to the slides for this investigation so will only need a relatively small volume of each solution.

For tips on calculating volumes for different concentrations see page 17. You should add an equal volume of mineral salt solution to each of the pollen grains to ensure the tube grows fully (a lack of mineral salt solution could be a factor that limits the growth of the pollen tube).

You should keep the microscope slides with the pollen on in a Petri dish with a lid. This will provide a humid atmosphere, which will help to ensure that the pollen does not dry out.

The dependent variable of your investigation is the length of pollen tube. To measure the pollen tube you will need to use the ×100 lens on the microscope and ensure that it has been calibrated (see pages 20–21 for more on this). It is important that you measure the pollen tube length quickly as the heat from the microscope lamp may dry out the samples, affecting the results.

You should repeat the investigation for each sucrose solution and then calculate a mean pollen tube length.

Focus on maths skills

Averages

Mean

The mean (often denoted by $\bar{x}$) is an average that is calculated by adding all the individual data values together and dividing by the total number of data points. It is the most commonly used average in biology and the one you will normally use in biological practical investigations.

The mean does have some disadvantages. It can be skewed by extreme results (outliers). If such values are present in a set of data, it might be appropriate to discard the outlier data before calculating the mean. It is important to think carefully before labelling a value as an outlier to ensure that results are not discarded unnecessarily.

For example, in Table 4 it seems clear that the mean for 0.6 has been adversely affected by what appears to be an outlier (trial 2). It would therefore seem sensible to remove this value when calculating the mean. This would then give a mean value of 250 for 0.6, which is more in keeping with the trend. The best option would be to repeat this trial and see if it delivered a less anomalous result. This could then be used when calculating the mean.

Table 4 Effect of sucrose concentration on pollen tube length

Sucrose concentration/ mol dm^{-3}	Pollen tube length/µm			
	Trials			
	1	2	3	Mean
0.0	65	79	71	72
0.2	215	201	232	216
0.4	253	101	247	200
0.6	265	289	232	262
0.8	110	99	102	104

Other types of average are outlined below.

Median

The median is the middle value in a data set. To find the median, arrange all the data points in order and pick out the middle value in the sequence. If there is an even number of data points, take the two in the middle and calculate their mean (i.e. add them together and divide by 2). The median is particularly useful in data with a large range where exceptionally high or low values (outliers) in the data would skew the mean.

Mode

The mode is the most common value in the data set. The mode can be used for non-numerical data or when the data points cannot be put in a linear order. If a data set does not contain any values that occur more than once, then it does not have a mode. This is not the same as saying the mode is zero — the mode would only be zero if 0 was the most common value in the data set.

Conclusions/evaluations

You will probably find that at high concentrations sucrose will become a limiting factor, potentially due to the osmotic effects of being a low-water-potential solution, and will give a lower pollen tube length.

This investigation often shows a great deal of variability in results due to a range of other factors affecting the pollen tube growth. If your results are showing an unexpected trend check with your teacher and other students in the class. As these are core practicals it is important that you not only know the procedures but also the expected results of the investigations.

Core practical 5

Investigate the effect of temperature on beetroot membrane permeability

CPAC 1a, 2a–2d, 4a, 4b, 5b

Background

Beetroot cells contain the red pigment betalain. The pigment is stored in the vacuoles of the cells. Betalain is too large a molecule to pass through the phospholipid bilayer of the tonoplast (membrane around the vacuole) or the external plasma membrane of the cell.

The plasma membrane or cell surface membrane is the barrier through which all matter entering cells must pass. It is selectively permeable. This means that it is able to control what enters and exits the cell.

The plasma membrane mainly consists of phospholipid molecules arranged into a bilayer (a double layer). Phospholipid molecules are made up of a hydrophilic glycerol phosphate 'head' and two fatty acid hydrophobic 'tails'. The hydrophilic heads of the phospholipid molecules face outwards and the hydrophobic tails face inwards. This forms a hydrophobic, non-polar region in the middle of the bilayer. This prevents polar molecules such as glucose from passing through.

Under the electron microscope and when stained with a water-soluble dye the cell membrane appears as a double line. This is because the hydrophilic heads of the phospholipid molecules take up the dye.

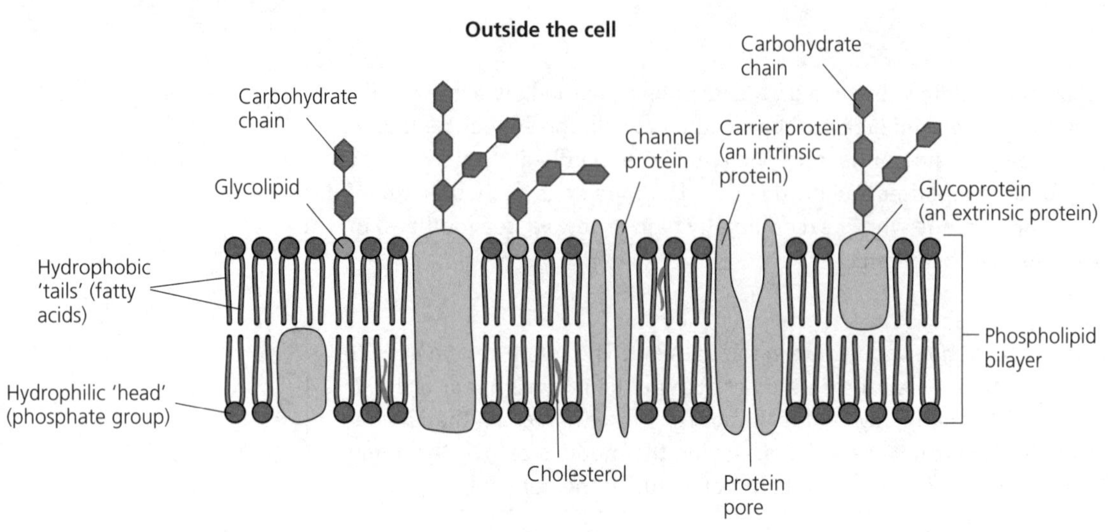

Figure 13 Structure of the plasma membrane

The structure of the plasma membrane is shown in Figure 13.

The plasma membrane contains proteins. These are divided into two types:

- Intrinsic proteins lie across both the layers of the membrane (e.g. a carrier protein).
- Extrinsic proteins are either in one layer of the membrane or on the surface of the membrane (e.g. certain enzymes found in the membrane).

When temperature increases, the molecules that make up the membrane gain kinetic energy and move at a faster rate. This causes the membrane to become more fluid (the components are more free to move) and therefore more permeable.

The plasma membrane is best described by the fluid mosaic model proposed by Singer and Nicholson:

- Fluid — all parts of the membrane can move relative to each other.
- Mosaic — proteins are dotted throughout the membrane like mosaic tiles.

In this investigation heating the cells causes the plasma membrane to become more fluid (the hydrophobic interactions between the phospholipid molecules weaken). This increases the permeability of the membrane, allowing betalain to leave the cells and diffuse into the surrounding solution. At high temperatures the proteins in the membrane will change shape, further disrupting the membrane structure and leading to an even greater release of betalain from the cells of the beetroot.

Practical notes

When beetroot is placed in solution the betalain will cause a colour change if it leaves the cells. Differing concentrations of betalain will lead to different intensities of colour change. These differences can be observed with the eye. However, to give a more accurate, quantitative value for the colour change, a colorimeter can be used.

The colorimeter should be set to measure the transmission of blue or green light. As the concentration of betalain in the solution increases, the colour will change to a darker pink or red colour. This will lead to a decrease in the transmission.

It is important when carrying out this investigation to ensure that you wash the beetroot discs thoroughly before placing them into the solutions. This ensures that damaged cells on the surface of the discs are removed — these would adversely affect the results by releasing betalain into the solution that is not related to the effect of temperature.

You should select a suitable range of temperatures between 0°C and 70°C that will give a good range of results. The membrane permeability should markedly increase above around 40°C, so you need to ensure that there is an adequate number of temperatures above this level to fully see the effect of this increase in permeability.

You will need to collect replicates (repeat readings at each temperature). The best way to do this would be to use a different sample of beetroot for each replicate. If this is not possible within the time constraints of your lesson you can use replicates from the same sample of beetroot. Once you have replicates at each temperature you should calculate the mean values for each temperature. These values can then be plotted on a graph. You should also plot range bars for each of your results.

Focus on maths skills

Range bars

Range bars can be plotted on graphs to show the spread of your results around the mean (Figure 14).

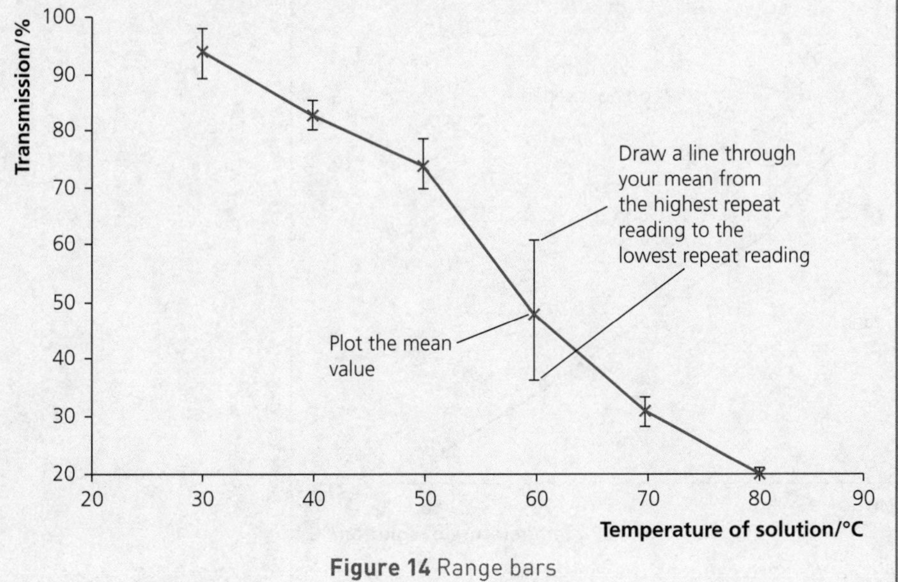

Figure 14 Range bars

It is important to ensure that the axes of your graphs allow you to plot range bars. For example, if the y-axis in Figure 14 was only plotted to 95 (understandable, as this was the mean value) then the range bar would not fit on.

Range bars allow you to make two different comments on your results:

- How **precise** the results are. This can be determined by the relative size of the range bars. If a range bar is relatively large this shows that at least one of the repeat results is not close to the mean. This would lead us to conclude that the results for this repeat are not precise. If a range bar is relatively small this shows that the repeat results are close to the mean, so the results at this temperature are precise.
  It is possible to have a mixture of results in an investigation that are precise and not precise. Also, you should not be worried if your results are not precise. This implies a flaw in the method that could be identified and improved. It is important in scientific research to fully report all results, even if they are not what was expected. This allows results to be evaluated and experimental processes to be improved, which should in turn improve the quality of the results.
- The **trend** shown by your results. You should look to see if your range bars overlap (see Figure 15 for an explanation of how to do this). If adjacent range bars overlap this shows a lack of confidence in the trend as it suggests the possibility that the variability of the repeats is responsible for the trend of the results. Non-overlapping of adjacent range bars increases the confidence in the trend (although it is no guarantee that it is correct). The one case where this concept does not apply is when a graph is reaching a plateau (Figure 16).

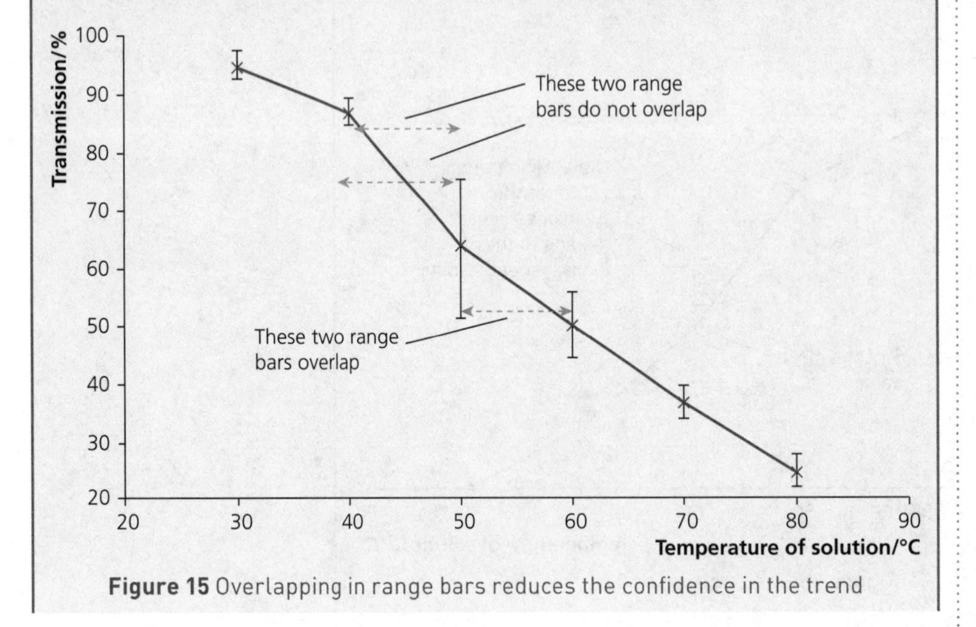

Figure 15 Overlapping in range bars reduces the confidence in the trend

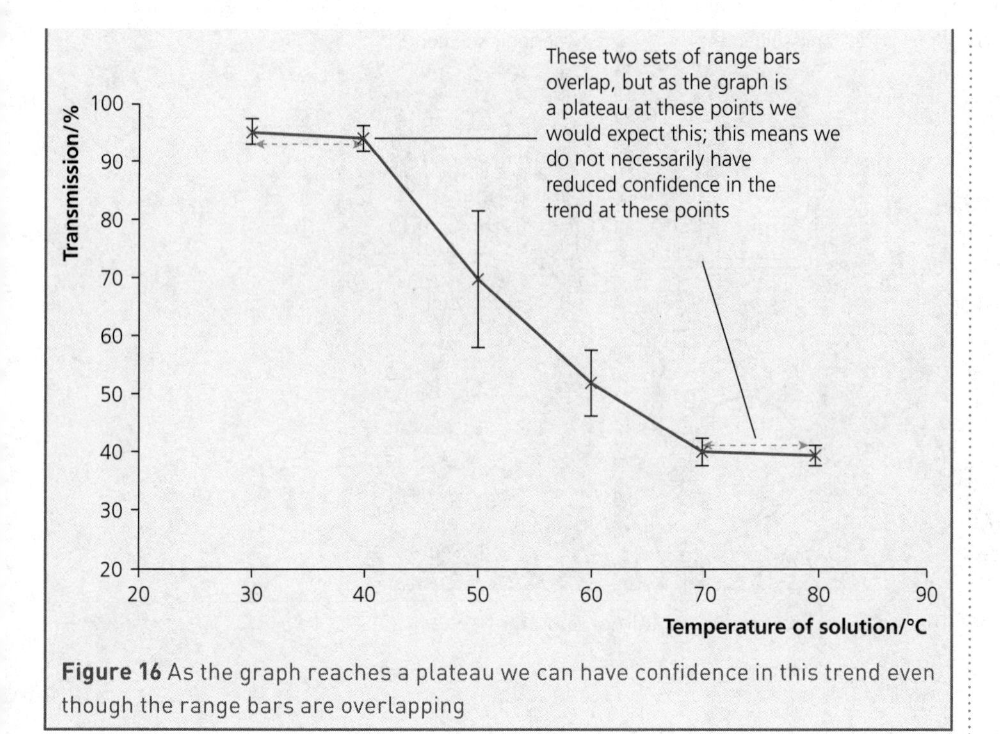

Figure 16 As the graph reaches a plateau we can have confidence in this trend even though the range bars are overlapping

Core practical 6

Determine the water potential of plant cells.

CPAC 1a, 2a, 2b, 4a, 4b

Background

In this investigation you are required to determine the water potential of plant cells.

Water moves across the plasma membrane by osmosis. Osmosis is the movement of water molecules from a high water potential to a lower water potential across a partially permeable membrane. Water potential is the potential energy of water relative to pure water.

A plant cell placed in a solution with a lower water potential than the cell will lose water and may become flaccid and plasmolysed. Plasmolysis is where the cytoplasm shrinks and comes away from the cell wall. If a plant cell is placed in a solution with a higher water potential than the cell it will gain water, swell and become turgid. An animal cell placed in a solution that has a higher water potential than the cell will burst as it does not have a cell wall. An animal cell in a solution that has a lower water potential than that of the cell may become shrivelled (Figure 17).

Water potential can be calculated using the following equation:

water potential of a cell Ψ = osmotic potential (π) + turgor pressure (P)

The osmotic potential (π) is generated by the solutes dissolved in the water.

The turgor pressure (P) is generated by the cytoplasm pushing on the cell wall. As the cell wall is rigid and inelastic it resists this pressure.

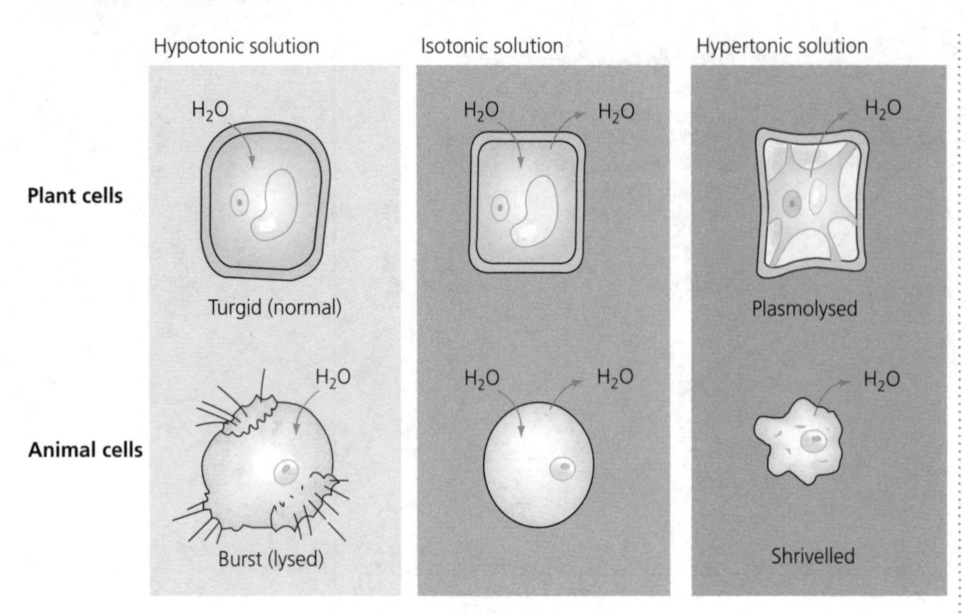

Figure 17 Effect of differing water potentials on animal and plant cells

Practical notes

In this investigation you will be using different concentrations of sucrose solution as the external solute. These sucrose solutions will all have different water potentials (Table 5).

Table 5 Water potentials of different sucrose solutions

Concentration of sucrose/M	Water potential of the solution/kPa
0	0
0.25	-680
0.5	-1450
0.75	-2370
1.0	-3510

The solutions with a lower water potential than the interior of the plant cells will cause water to leave the cells by osmosis, leading to a decrease in mass of the plant tissue. The solutions that have a higher water potential than the interior of the cell will cause water to enter the cell by osmosis, leading to an increase in the mass of the plant tissue.

When the water potential of the solution is equal to the water potential of the cells then there will be no net movement of water, and hence no change in the mass of the plant tissue.

By determining the water potential of the solution at the point where there is no change in mass you will then know the water potential of the cells in the plant tissue sample.

You should ensure that the tissue samples have the same lengths, widths and diameters. It will be difficult, however, to ensure that all the samples have the same mass. As each tissue sample may be slightly different in mass, rather than using absolute change in mass as your dependent variable you should use % change in mass. This allows data from tissues with different masses to be compared.

Focus on maths skills

Percentage change

Like fractions, percentages are part of a whole, but they are expressed in the form of a number followed by the percentage symbol %, which means 'divided by 100' or 'out of 100'.

To calculate percentage change, first find the change. For example, if the mass of a potato sample changed from 10 g to 8 g then the change is:

$8 - 10 = -2\text{ g}$

Now divide this change by the original mass and multiply by 100:

$\frac{-2}{10} \times 100 = -20\%$

change in mass = −20%

You will need to pick a range of different concentrations of sucrose solution. It is unlikely that any of your concentrations will actually give no change in mass. By using a wide range of sucrose solutions some will cause a decrease in mass and some will cause an increase. By plotting these data on a graph the point of no change in mass can be determined by finding the intercept point on the *x*-axis.

Focus on maths skills

Determining the intercept of a graph

The intercept of a graph is the point where the graph crosses one of the axes. In exam questions you would normally be asked to find the intercept on the *x*-axis. To do this, either read off the *x* value at which the line or curve crosses the *x*-axis or, if the crossing point is not shown on the graph, you may be able to extrapolate from the line to find where it intersects the axis. This is shown in Figure 18. Here the intercept of the graph is approximately −1800 kPa. This means that the concentration of sucrose solution that would be predicted to give no change in mass is −1800 kPa.

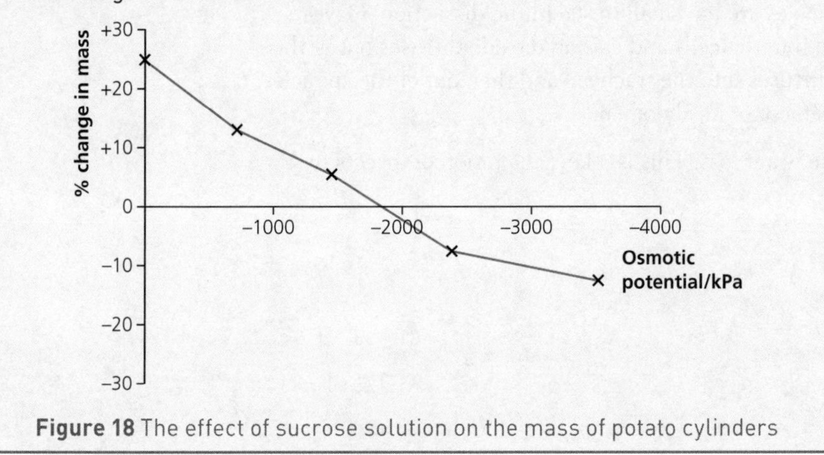

Figure 18 The effect of sucrose solution on the mass of potato cylinders

Extension

An important follow-up investigation would be to make up a solution of the concentration that was predicted to give no change in mass and test out this assertion. If this concentration of sucrose solution does cause a change in mass, a small range of concentrations around it could be tested to more accurately pinpoint the exact concentration that gives no change in mass.

The investigation could also be repeated with different plant tissues to study the range of different water potentials. This investigation could not be extended to animal tissues due to their lack of cell walls and tendency to burst in solutions that have a higher water potential than the interior of the cells.

Core practical 7

Investigate the gas exchange system of a locust

CPAC 1a, 2a, 2b, 3a–3c, 4a

Background

This practical deals with dissecting insects. It is likely that you will be using locusts. Locusts are invertebrates and display a very different morphology and physiology from a mammal such as a human. Locusts are arthropods so have an exoskeleton (exterior skeleton) made from the polysaccharide chitin. They also have segmented bodies and segmented legs.

As locusts are insects they have wings and three pairs of segmented legs. Like all insects they have an open circulatory system where blood (haemolymph) is pumped around the body cavity by a simple tubular heart. Insects do not possess a respiratory pigment (e.g. haemoglobin in humans) and oxygen is not carried by haemolymph.

Insects instead have a tracheal gas exchange system that carries oxygen directly to all parts of the body (Figure 19). The gases enter and leave through small holes in the abdomen called spiracles. The gases then diffuse through tubes supported by chitin called tracheae. The chitin ensures that the tracheae do not collapse due to pressure changes. The tracheae then subdivide to form smaller tracheoles, which lead directly to the insect's tissues. The tracheoles are too small to see in the dissection. Oxygen diffuses from the tracheoles and into the cells and carbon dioxide diffuses out of the cells. The carbon dioxide then diffuses into the tracheae and then out of the spiracles. The insect can ventilate by movement of its abdomen.

The spiracles can close to reduce water loss. This is a key adaptation of insects to terrestrial life.

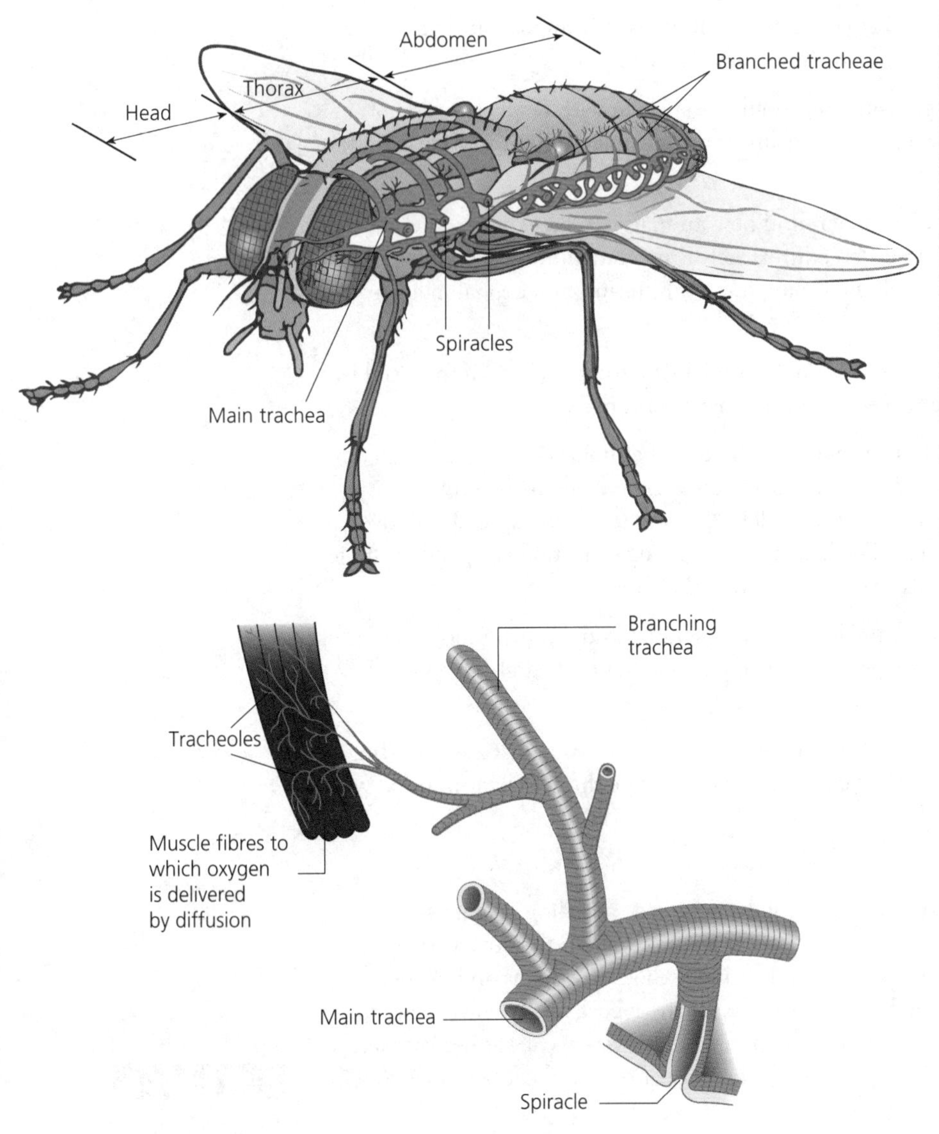

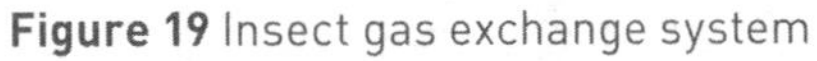
Figure 19 Insect gas exchange system

Practical notes

An important element of this practical is the ethical issue raised by the use of animals. Use of animals in science is a contentious issue and there is a wide range of different viewpoints. A positive reason why locusts should be killed for this investigation is that the benefits derived from it outweigh the harm caused. Some of the main benefits from dissections are outlined below:

- Knowledge of whole organisms is advanced in a way that cannot be done without dissections.
- The true nature of fresh tissue can be observed and studied and the variation that can exist across organisms can also be seen.
- Manual skills are developed.

These issues need to be balanced against the harm done. It is important that you derive the maximum benefit from any dissection investigation.

When dealing with living organisms you should treat them with respect — do not subject them to unnecessary pain or stress and generally ensure their welfare as much as possible.

There is also a regulatory framework around how animals can be experimented on, based on the Animals (Scientific Procedures) Act. Protected animals are all living vertebrates (except humans), including some immature forms, and cephalopods (e.g. octopus, squid, cuttlefish).

The guiding principle for the use of animals in scientific work can be summarised by the 'three Rs' — **replacement**, **reduction** and **refinement**.

Replacement involves avoiding or replacing the use of animals defined as 'protected' under the Animals (Scientific Procedures) Act. This can include using human volunteers, mathematical and computer models, cell and tissue lines, or alternative organisms such as invertebrates. The locust dissection investigation is a good example of replacement as invertebrates are used as opposed to frogs or mice.

Reduction aims to use fewer animals per experiment or study. You can do this in investigations by working in groups and ensuring that you are as careful as possible not to waste any specimens.

Refinement refers to methods that minimise the pain, suffering, distress or lasting harm that may be experienced by the animals. This is something to bear in mind with all use of animal samples

Dissection

When you use a microscope you are essentially looking at two-dimensional sections through plant and animal organs. Much of what we know about the structure of animals and plants comes from dissection, in which whole organisms and/or their organs are cut open or sectioned. When you dissect the locust you are investigating the relationships between organs and you can make observations about their structure and arrangement within the body. To carry out a dissection you must be provided with dissection instruments, which may include:

- at least one sharp scalpel
- a pair of sharp, fine scissors
- a pair of coarse forceps
- a pair of fine forceps
- at least one mounted needle
- a blunt seeker

You also need a hand lens (×10) and/or a stereomicroscope. Handling these instruments to expose the internal structures of organisms is one of the practical skills that you need to demonstrate.

Always follow instructions about safe disposal of dissected material. Place the dissecting instruments that you have used into a solution of a disinfectant.

Practical tip

Take photographs of each stage of your dissection. You can use them to make good revision notes by writing about how the structures you see are adapted to carry out their specific functions of gas exchange.

Dissection safety

- Check to make sure that you do not have any allergies that might prevent you carrying out the dissection.
- Wear a lab coat and if you have any cuts or lesions on your hands wear gloves.
- Make sure that you have a suitable surface on which to dissect, such as a cork mat or a special dissecting dish containing hard wax if you need to hold the specimen with pins. Absorbent benchcoat, wooden boards or plastic trays are suitable if pins are not necessary.
- When carrying out dissections you should always ensure that you are cutting away from your body. Keep sharp instruments somewhere safe on the bench so that they are not inadvertently dropped on the floor.
- Use sharp equipment; blunt equipment increases the risk of slipping and cutting yourself.
- Use the appropriate implement for a task — for example, often small dissection scissors can be more useful than a scalpel. This includes not trying to cut through structures that are too solid, such as using a small scalpel to cut through a piece of bone.

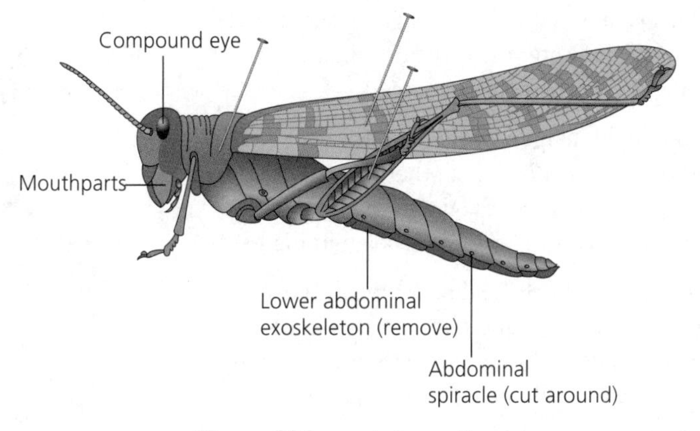

Figure 20 Locust dissection

Figure 20 shows a locust dissection. The procedure is as follows:

1. Use a hand lens or dissection microscope to find the spiracles on the thorax and abdomen of a dead locust.
2. Place the dead locust into a dissection dish on its ventral surface. Cut off the wings and hold it in place with pins positioned either side of the thorax.
3. Pull the abdomen back and push a pin through the end of the abdomen and into the wax to hold it in place. Use a pair of fine scissors to cut into the exoskeleton at the top of one side of the abdomen. Cut along the edge towards the thorax and towards the head.
4. Make another cut alongside the other side of the abdomen and use a scalpel to separate the exoskeleton from the internal organs underneath.
5. Cover the dissection in water.
6. The tracheal system consists of many large tracheae that are easy to see as they are like silver threads. Use forceps and a blunt seeker to trace the system throughout the body. Look for air sacs in the thorax.

Practical tip

Search online for electron micrographs of tracheoles, which form the gas exchange surface in insects. You will not see them in your dissection.

7 Remove a small piece of a trachea where it divides. Make a temporary preparation in water on a microscope slide. Cover with a coverslip and observe under low and medium power. You will see the rings of chitin that keep the tracheae open to keep a low resistance to the flow of air.

Core practical 8

Investigate the effect of environmental conditions on water uptake in a plant shoot

CPAC 1a, 2a–2d, 4a, 4b, 5b

Background

This practical involves investigating the effect of a factor on the uptake of water by a plant shoot.

Plants draw water up through the xylem. This is due to the **transpiration stream** and cohesion forces between water molecules (the cohesion–tension theory).

Water is known as a **polar molecule**. This is due to it having a slightly uneven distribution of charge. The hydrogen atoms of the water molecule are slightly positively charged while the oxygen atom is slightly negatively charged.

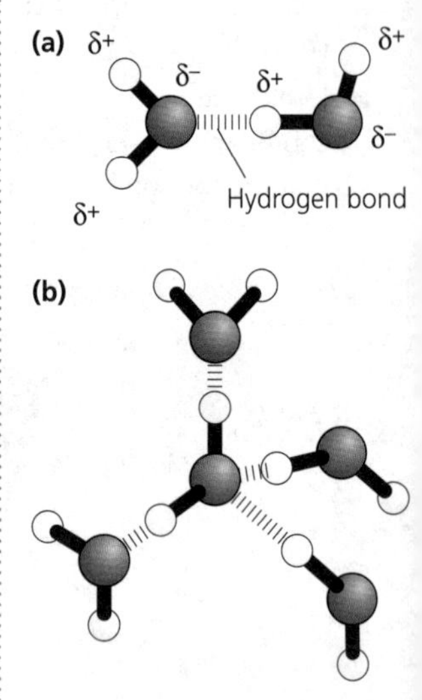

Figure 21 Hydrogen bonding between water molecules

This slight uneven distribution of charge allows **hydrogen bonds** to form between the hydrogen atom of one water molecule and the oxygen atom of another water molecule. It is these hydrogen bonds that create the force known as cohesion that 'sticks' the water molecules together. Hydrogen bonding between water molecules is shown in Figure 21.

Water evaporates from the surface of the mesophyll cells and out through the stomata on the leaves of the plants. These water molecules are joined by cohesion forces to other water molecules, forming an unbroken column through the xylem down into the roots. The evaporation of the water molecules 'pulls' on this column of water molecules, exerting 'tension' which draws the water molecules up the xylem.

Practical notes

Any factor that changes the rate of transpiration of water from the leaf will also affect the rate of water uptake by the plant shoot. The key factors that affect the rate of transpiration are:

- temperature
- light intensity
- air movement
- humidity

Increasing the **temperature** of the environment will increase the kinetic energy of the water molecules, increasing the rate of evaporation of water.

Increasing **light intensity** will lead to an increase in the rate of photosynthesis. This will then lead to an increase in stomatal opening to allow the gas exchange required for photosynthesis to occur. The increase in stomatal opening leads to a greater rate of transpiration.

As water evaporates from the inner surfaces of the leaves and then diffuses through the stomata it can create a humid microclimate around the stomatal opening. This leads to a reduction in the diffusion gradient between the inside of the stomata and the outside of the stomata, reducing the rate of water loss. An increase in **air movement** will lead to the humid air around the stomatal opening being removed and being replaced with drier air, leading to a greater diffusion gradient and therefore increase in the rate of transpiration.

Humidity is a measure of the concentration of water in the air. A higher humidity level will decrease the concentration gradient between the exterior air and the air inside the leaf. This will then lead to a decrease in the concentration gradient between the inside of the leaf and the outside of the leaf, lowering the rate of transpiration. A low humidity would lead to a greater concentration gradient and so an increase in the rate of transpiration.

Only one of these will be used as the independent variable in your investigation. As they all affect the rate of transpiration you will need to control all the factors that you are not using as independent variables.

Two other potential variables are the leaf surface area and the density of stomata on the leaf (how many are present per unit area). This is because more stomata will lead to an increase in the rate of transpiration. These two factors can both be controlled by ensuring the same shoot is used for each of the repeats of the investigation. The surface area of the underside of leaves can be measured and used to calculate the volume of water taken (cm^3) per area of leaf (cm^2).

Practical tip

Humidity is the only one of the key factors that decreases the rate of transpiration as it increases. All the others increase the rate of transpiration as they increase.

Figure 22 shows a potometer set up to take readings.

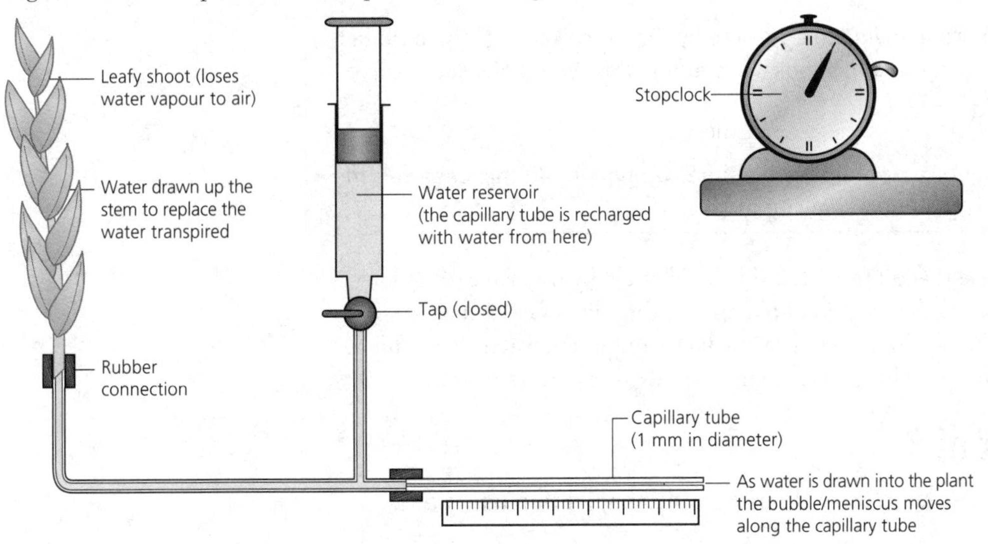

Figure 22 A potometer

There are some important set-up points to note:

- It is important to cut the shoot under water to ensure no bubbles enter the xylem and disrupt the transpiration stream, which relies on the fact that the water molecules are all joined to each other by cohesive forces. For the same reason it is important to ensure all the links in the apparatus are airtight. This can be done by making sure that all the links are fully attached and by smearing petroleum jelly around them.

- The leaves must be kept dry.
- The apparatus must be assembled under water and only one bubble allowed to enter the tube of the potometer.
- The apparatus should then be removed from the water and the reservoir filled; the reservoir is used to set the bubble to a known position. By opening the tap on the reservoir, water enters the main tube and causes the bubble to move to the right. It is important to only open the tap a fraction as if too much water enters the tube from the reservoir it will push the bubble out of the end of the tube.
- The distance moved by the bubble in a time can then be used to calculate the rate of water taken up. First calculate the volume of water using the formula for the volume of a cylinder (as the water moved along a cylindrical tube). In this case the capillary tube is 1.0 mm in diameter and the bubble has moved 4.3 cm (43 mm).

Focus on maths skills

Potometer calculations

The volume of a cylinder is worked out as follows:

volume of cylinder = area of one circular end × length

area of a circular end = πr^2

$$r = \frac{\text{diameter}}{2} = \frac{1.0}{2} = 0.5$$

area = $\pi \times 0.5^2 = 0.79\,\text{mm}^2$

volume = $0.79 \times 43 = 34\,\text{mm}^3$

Now to calculate the rate, divide this volume by the time taken for the bubble to move the distance. Assuming this took 11 minutes, this would give a rate of:

$$\frac{34}{11} = 3.1\,\text{mm}^3\,\text{min}^{-1}$$

Units for this rate require a volume and a time component. In this case volume is in mm^3 and time is in minutes, giving $\text{mm}^3\,\text{min}^{-1}$.

It is important to realise that all the water that is taken up by the plant is not lost through transpiration. Water is required to maintain turgidity of cells and as a reactant in photosynthesis, although only a small amount of the water drawn up by the xylem is actually used in this process. The majority is lost in transpiration.

Core practical 9

Investigate factors affecting the rate of aerobic respiration using a respirometer

CPAC 1a, 2a, 2b, 4a, 4b

Background

You will be investigating the rate of respiration in a small organism.

This investigation will involve the use of a piece of apparatus known as a respirometer. There are many different designs of respirometer; an example is shown in Figure 23.

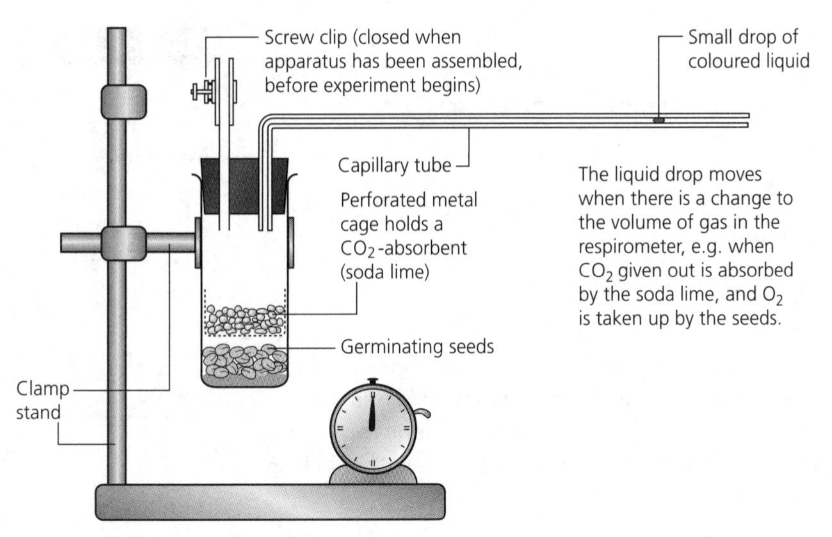

Figure 23 One design of respirometer

The organisms are placed in a sealed tube. If they are respiring aerobically they take in oxygen and release carbon dioxide:

$$C_6H_{12}O_6 + 6O_2 \rightarrow 6CO_2 + 6H_2O$$

The carbon dioxide is formed by the decarboxylation of respiratory substrates in the link reaction and Krebs cycle. Oxygen acts as the final electron acceptor in the electron transport chain during oxidative phosphorylation.

The respiration equation shows that the volume of oxygen taken in should be equal to the volume of carbon dioxide released. This would make it challenging to measure the rate of respiration as the total volume of gas in the tube wouldn't change! To overcome this problem respirometers use a method of absorbing carbon dioxide. This is usually done by using soda lime or limewater in the apparatus.

As the tube is sealed, no gas can enter and the carbon dioxide is absorbed by the soda lime. This decreases the total volume of gas in the tube containing the organisms. As the volume of gas in the tube decreases the pressure in the tube also decreases. This leads to a pressure differential that draws liquid along the tube towards the organisms. By measuring the distance moved by the liquid over time the rate of respiration can be determined.

A more advanced method could be to use oxygen and carbon dioxide detectors linked to a data logger. The change in concentration of these two gases as the organisms respired could then be determined. This method would remove the need for the soda lime and the coloured liquid.

Practical notes

You should measure the distance moved by the liquid in a set time. If the liquid moves 24 mm in 5 minutes:

$$\text{distance per minute} = \frac{24}{5} = 4.8\,\text{mm min}^{-1}$$

It is important to apply the principles of ethical use of organisms to this investigation. These are detailed fully on pages 39–40.

You should measure the mass of the organisms before they are placed in the respirometer. The mass of an organism should have an effect on the rate of respiration occurring, even across organisms of the same species. For this reason the rate of respiration should be determined per gram, as follows:

total mass of germinating seeds = 3.5 g

rate of liquid movement = $4.8\,\text{mm min}^{-1}$

rate of movement per gram = $\frac{4.8}{3.5} = 1.4\,\text{mm min}^{-1}\,\text{g}^{-1}$

This investigation does rely on some key assumptions, i.e. that the change in gas volume is being caused by respiration of the organisms and the carbon dioxide is being absorbed as expected. To test this a control can be used without any soda lime but with all other variables the same. As the volume of oxygen taken in will be the same as the volume of carbon dioxide released in the control investigation there should be no movement of the liquid in the capillary tube.

An additional control can also be set up with an inert material of approximately the same volume as the organisms in the experimental respirometer but with all other conditions the same. Again there should be no movement of liquid in the capillary tube. If the coloured liquid is moving in the control tube, wait for it to stop before starting to take readings with the experimental apparatus.

As this investigation involves living organisms the data may have quite a spread. To get an idea of this spread or dispersion of the data the range and standard deviation can be calculated.

Practical tip

You can also calculate the volume of oxygen absorbed by using the formula for volume of a cylinder (page 44). You will need the distance moved by the coloured liquid and the diameter of the tube it is moving through.

Focus on maths skills

Measures of dispersion

There are two measures of dispersion that you can use here:

- **Range** — the difference between the largest and smallest data values. A larger range indicates a greater spread of data.
- **Standard deviation** — the square root of the variance s^2. Variance is the sum of the squared deviations of the data values from the mean value, divided by the number of data values.

Below is a worked example of calculating the range and standard deviation of some data from this investigation.

Table 6 Respirometer investigation results

Rate of fluid movement per gram/mm min^{-1} g^{-1}
1.10
1.20
1.80
0.90
1.20

range = largest value − smallest value

= 1.80 − 0.90

= 0.90 mm min^{-1} g^{-1}

It is often easier to calculate the standard deviation using a table.

Step 1: find the mean value:

$$\bar{x} = \frac{1.10 + 1.20 + 1.80 + 0.90 + 1.20}{5} = 1.24$$

Step 2: find the difference between each data value and the mean, $x - \bar{x}$. Then square each of these differences to get $(x - \bar{x})^2$.

Table 7 Deviations from the mean

Rate of fluid movement per gram/mm min^{-1} g^{-1}	$x - \bar{x}$	$(x - \bar{x})^2$
1.10	−0.14	0.0196
1.20	−0.04	0.0016
1.80	0.56	0.3136
0.90	−0.34	0.1156
1.20	−0.04	0.0016

A good way of checking if the deviations from the mean are correct is to sum them. This sum should always equal zero.

Step 3: find the average squared difference, i.e. the variance s^2, by adding up the squared deviations and dividing the sum by the total number of data points.

$$\sum(x - \bar{x})^2 = 0.452$$

$$s^2 = \frac{\sum(x - \bar{x})^2}{5} = \frac{0.452}{5} = 0.0904$$

Step 4: calculate the square root of the variance to find the standard deviation:

$$s = \sqrt{0.0904} = 0.3$$

A standard deviation closer to zero indicates that the data points do not deviate much from the mean, which suggests that the data are more consistent. In this example, the standard deviation is relatively small, because there is little deviation from the mean. The value for the range is also fairly low. We can therefore conclude that there is not a large degree of variation in the data and that they are precise (the replicate results are close together).

Extension

Further analysis of the results can be done to calculate the volume of oxygen taken up. To do this, find the diameter of the tube containing the coloured liquid, then use the equation for the volume of a cylinder, $\pi r^2 h$, where h is the distance travelled by the liquid, to calculate the volume of oxygen taken up.

To improve the accuracy of the results a gas syringe could be used to measure the volume of gas rather than the tube containing the coloured liquid.

This investigation can be extended by studying the effect of changing temperature on the rate of respiration of the organisms. As respiration is an enzyme-controlled process, the rate of respiration should show a marked increase as temperature increases. If temperature is being changed it becomes even more important to consider the ethical treatment of organisms.

Anaerobic respiration by yeast can also be investigated using a respirometer. Soda lime should not be used. The yeast suspension should be placed at the bottom of the boiling tube with another smaller tube entering the suspension and connected directly to the gas syringe or tube containing coloured liquid. A layer of oil should be placed over the yeast. This will prevent any oxygen entering the suspension. The yeast will quickly use up the oxygen present in the suspension and begin respiring anaerobically.

The yeast cells will release carbon dioxide due to the decarboxylation of pyruvate to ethanal during anaerobic respiration. This carbon dioxide will then be collected by the gas syringe or cause the coloured liquid to move to the right.

Core practical 10

Investigate the effects of different wavelengths of light on the rate of photosynthesis

CPAC 1a, 2a–2c, 4a, 4b, 5b

Background

You will be investigating the effect of different wavelengths of light on the rate of photosynthesis.

Different photosynthetic pigments absorb light of different wavelengths. The absorption spectrum (Figure 24a) shows the absorption of different wavelengths of light by the leaf. The action spectrum (Figure 24b) shows the rate of photosynthesis at different wavelengths. The fact that its shape matches that of the absorption spectrum shows that the wavelengths of light absorbed by the photosynthetic pigments are those used in the biochemical reactions to produce the photosynthetic products.

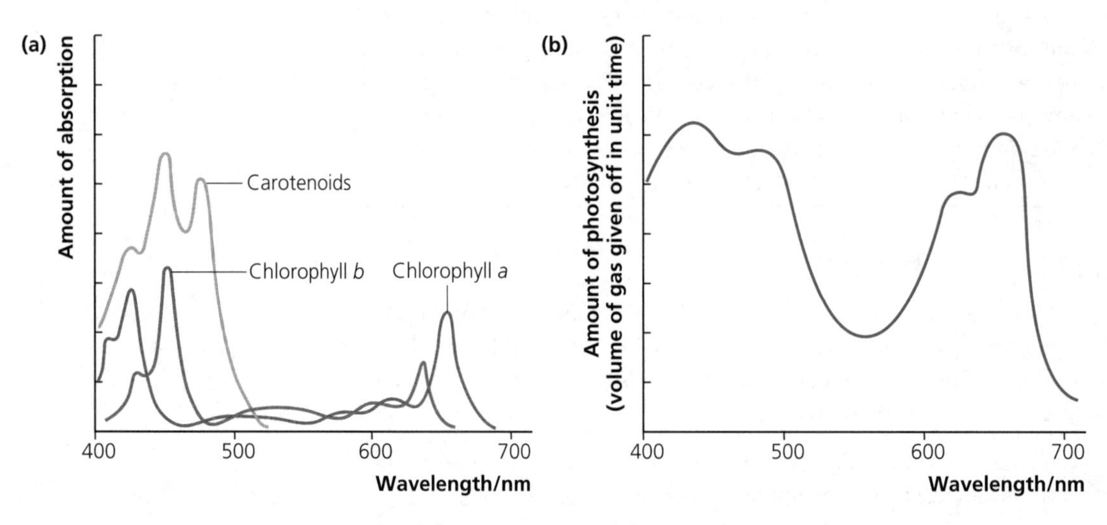

Figure 24 (a) Absorption and (b) action spectra

The photosynthesis equation is:

$$6CO_2 + 6H_2O \rightarrow C_6H_{12}O_6 + 6O_2$$

To measure the rate of photosynthesis, the rate at which products are formed or reactants used up can be measured. The most common method is to measure the rate of oxygen production. Oxygen is produced by the photolysis of water during non-cyclic photophosphorylation in the light-dependent stage.

The rate of carbon dioxide uptake can also be measured. This is often done using aquatic plants, where an indicator shows the change in pH of the water as the dissolved carbon dioxide decreases. The rate of formation of photosynthetic products (e.g. starch) is not normally measured in A-level practicals, but there are methods of carrying this out. Carbon dioxide is taken in and the photosynthetic products are formed during the Calvin cycle in the light-independent reaction.

Practical notes

The simplest method of determining the rate of photosynthesis is to measure the volume of oxygen produced over time. The oxygen is released during photolysis in non-cyclic photophosphorylation. A greater rate of photosynthesis will lead to more oxygen being released per unit time. By using a plant that lives in water the gas produced will bubble up through the liquid and can then be easily collected and its volume read off. Figure 25 shows the set-up of the apparatus.

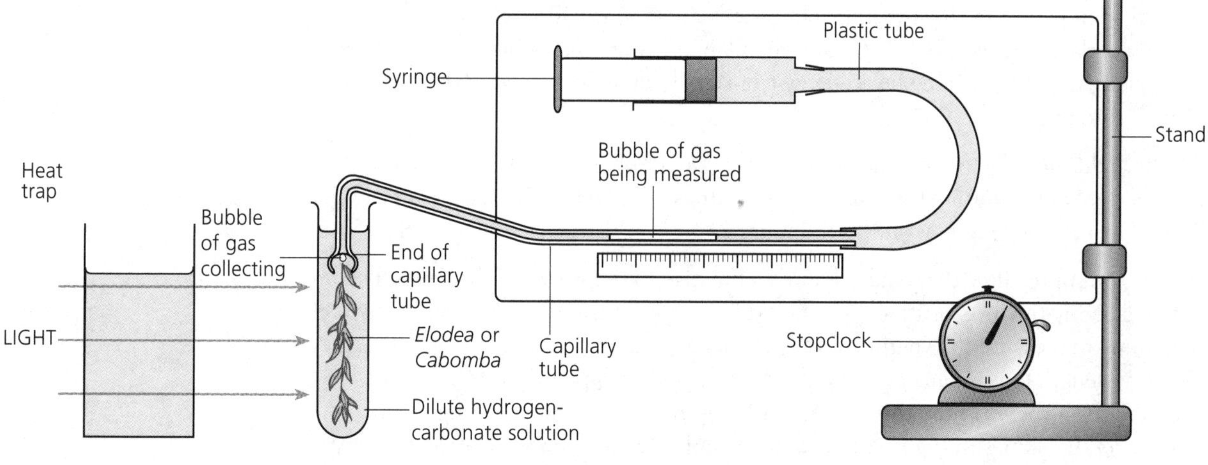

Figure 25 Measuring the rate of photosynthesis

The different wavelengths of light can be achieved using coloured filters. Each filter absorbs all colours of light except for the one it transmits. Using coloured filters that transmit known wavelengths of light allows you to plot your results as a line graph of rate of photosynthesis over wavelength.

There are a number of different limiting factors that affect photosynthesis, which need to be controlled in this experiment:

- **Light intensity** — in general, increasing light intensity will increase the rate of photosynthesis if other factors are not limiting. The most common way to control light intensity in a school or college photosynthesis investigation is to position the lamp the same distance from the plant for each experiment. This will not completely ensure the light intensity is always the same, so a light intensity meter could be used to give a more precise reading. There can also be issues with the colour filters used in the investigation because a darker filter can reduce the light intensity. Again, a light meter could be used to ensure that the same intensity of light is reaching the plant through each of the filters.
- **Carbon dioxide concentration** — carbon dioxide is a reactant in photosynthesis so its concentration can be a limiting factor if too low. To ensure there is an excess of carbon dioxide available to the plant, dissolve a known mass of carbonate (e.g. sodium hydrogen carbonate) into the water. This ensures that there is adequate carbon dioxide in the water. It is important, however, to make sure that you do not add too much carbonate.
- **Temperature** — temperature does not quite have the same effect on photosynthesis as it does in respiration, due to the fact that the first stage of photosynthesis is more photochemical than enzyme-controlled; nevertheless, it can affect the rate so needs to be controlled. Variations in room temperature should not have a significant effect, but what could be more important is the heat generated by the lamp used in the investigation. A heat shield in the form of a beaker of water can be placed between the lamp and the plant. As water has a high specific heat capacity it will absorb the heat energy from the lamp while allowing the light to pass through.

All these different factors interact to affect the rate of photosynthesis. When one factor is no longer limiting the rate of photosynthesis, the rate will increase until it is limited by another factor, as shown in Figure 26.

It is important to allow the pondweed to acclimatise to the wavelength of light used before starting the investigation. You should leave it a set time (e.g. 5 minutes) before beginning to measure the volume of gas produced. This will help ensure it is the new wavelength that is causing the gas to be released as opposed to the previous wavelength. If there is time this can be achieved by calculating a running mean. Take several readings in the first few minutes and calculate a new mean after every one. Once these mean values begin to be close together this suggests that the rate of photosynthesis has stabilised.

A possible disadvantage of this investigation is not knowing if the gas bubbles being produced are oxygen. If the plant is respiring at a greater rate than it is photosynthesising, it could be carbon dioxide that is being released. The point at which the plant switches from net uptake of oxygen to net release of oxygen is known as the compensation point. By controlling the temperature and the light intensity this should ensure that the plant is photosynthesising enough to be releasing oxygen.

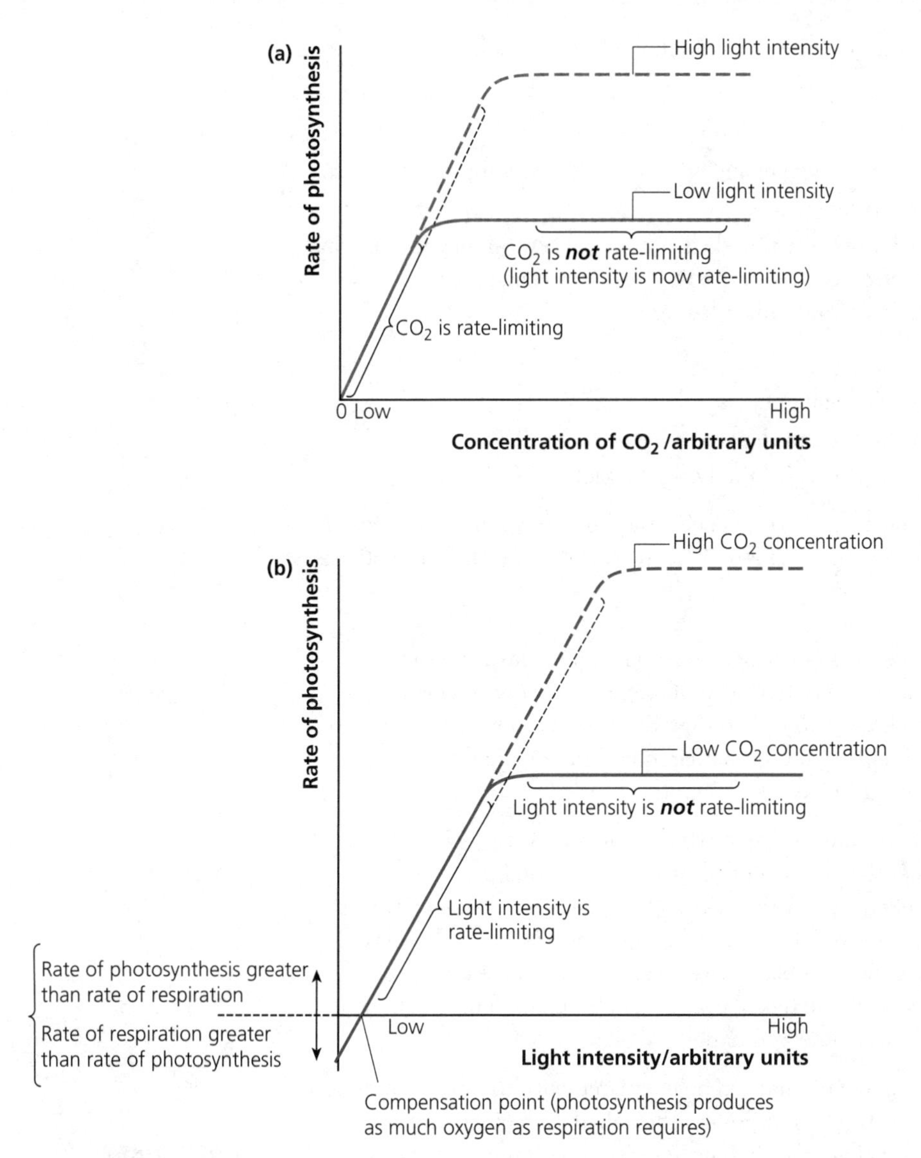

Figure 26 (a) Carbon dioxide concentration and (b) light intensity as limiting factors

Core practical 11

Investigate the presence of different chloroplast pigments using chromatography

CPAC 1a, 2a, 2b, 3a, 3b, 4a, 4b, 5b

Background

Within chloroplasts are a range of different pigments located on the membranes of the thylakoids and grana. They absorb light energy for use in the light-dependent reaction of photosynthesis. This energy is used to produce ATP and reduced NADP. These products are then transferred to the light-independent reaction in the stroma. The light-independent reaction then synthesises the complex organic molecules that are the final products of photosynthesis.

In flowering plants, there are two major groups of chloroplast pigments — chlorophylls and carotenoids.

Chlorophylls

- Chlorophyll *a* is the most common pigment and is found in all photosynthetic organisms.
- Chlorophyll *b* is found in flowering plants.
- Phaeophytin is a chlorophyll molecule that does not have a central magnesium ion. Phaeophytin is the first electron carrier in the PSII electron transport chain and is the photosynthetic pigment in purple sulfur bacteria.

Carotenoids

- Carotenes (α and β) are orange pigments.
- Lycopene is a bright pigment found in tomatoes.
- Xanthophylls include lutein and zeaxanthin, both of which are yellow.

By having a range of different photosynthetic pigments the plant maximises the range of wavelengths of light that the chloroplast can absorb, increasing the rate of photosynthesis.

Practical notes

Chromatography is a technique for separating, identifying and measuring the quantities of substances extracted from biological material. The substances being analysed are extracted in a specific solvent and then drops of the solution are placed at the start line at the base of a piece of chromatography paper or a layer of fine-grained material, such as silica gel, on a solid support.

The paper or thin-layer is placed into a solvent, which moves upwards. The solvent molecules are carried different distances depending on their solubility in the solvent and their attractions to the solid phase. When the solvent has reached about 10 mm from the end, the paper or plate is removed and a line drawn to indicate the solvent front. Pigments, such as those in chloroplasts, are visible. Other substances, such as amino acids, are not visible with the naked eye so the chromatograms might be sprayed with a special chemical or observed under UV light.

The distances moved by the pigments can then be compared with the distance moved by the solvent itself (the solvent front). This allows the calculation of a retention factor (Rf) value. The experiment Rf values can then be compared with reference values and the pigments within the initial mixture identified (Figure 27).

$$\text{Rf} = \frac{\text{distance travelled by substance}}{\text{distance travelled by solvent front}}$$

For example, a pigment moves 7.1 cm and the solvent front moves 9.8 cm:

$$\text{Rf} = \frac{7.1}{9.8}$$

$$\text{Rf} = 0.72$$

Note that Rf values are dependent on the solvent and solid used. This means that, potentially, the Rf values in the reference table may not exactly match those obtained in your investigation.

Practical tip

The retention factors, like other ratios, do not have units. They are expressed as a single number. A substance insoluble in the solvent has an Rf value of 0. One that is completely soluble has an Rf value of 1.

The Rf values of some of the common pigments that can be found in the mixture extracted from a leaf are shown in Table 8.

Table 8 Rf values of photosynthetic pigments

Pigment	Colour of spot	Rf value
Carotene	Yellow-orange	0.95
Phaeophytin	Grey-yellow	0.83
Xanthophyll	Yellow-brown	0.71
Chlorophyll *a*	Blue-green	0.65
Chlorophyll *b*	Light green	0.45

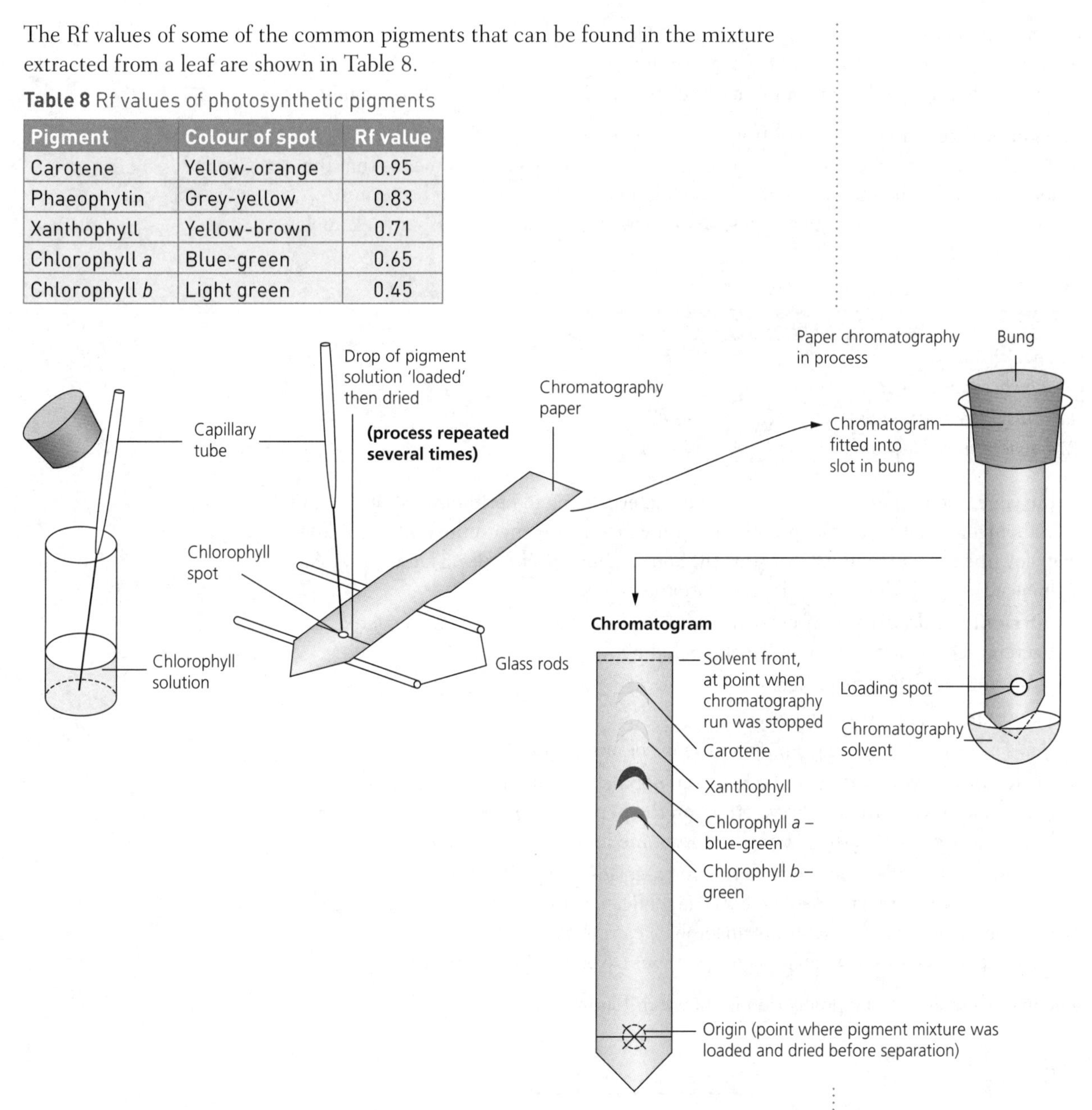

Figure 27 Chromatographic separation of pigments

Core practical 12

Investigate the rate of growth of bacteria in liquid culture

CPAC 1a, 2a, 2b, 3a, 3b, 4a, 4b, 5b

Background

The number of bacterial cells in a liquid culture can be estimated by the turbidity of the sample. The more turbid the sample the more bacteria present.

Bacteria are prokaryotes and divide by binary fission. This is an asexual process where a bacterium splits to form two identical bacteria. It is different from mitosis in a number of ways, including spindle apparatus not developing and the lack of a nuclear membrane.

The **generation time** is the time from a new bacterium being formed to it undergoing binary fission and a new generation being produced. This means that the generation time is the time taken for the population to double (as each bacterium divides to form two). Generation times vary depending on the bacteria; some examples are shown in Table 9.

Table 9 Bacterial generation times

Species of bacterium	Generation time/minutes
Escherichia coli	17
Staphylococcus aureus	27–30
Lactobacillus acidophilus	66–87
Mycobacterium tuberculosis	792–932

These generation times are all based on the bacteria being in optimum conditions and there being no limiting factors present. Limiting factors are environmental conditions that slow the rate of a population's growth. Some examples of limiting factors that can influence the growth of a bacterial population include:

- **nutrient availability** — if nutrients in the culture medium become depleted. Bacteria require a variety of nutrients including:
  - a source of carbon (e.g. glucose)
  - a source of energy (e.g. glucose)
  - a source of nitrogen (e.g. ammonium ions or amino acids)
- **build-up of toxic waste** — the bacterial population will produce waste that if not removed will have a negative effect on the population growth, for example by leading to a change in the pH of the culture medium.
- **oxygen** — dependent on the bacteria grown, a lack of oxygen or an increase in oxygen can limit growth. Aerobic bacteria (obligate aerobes) require oxygen to grow, while anaerobic bacteria (obligate anaerobes) cannot grow in the presence of oxygen. Facultative anaerobes are able to grow without oxygen but grow best in oxygen.

The growth of a bacterial population is shown in Figure 28.

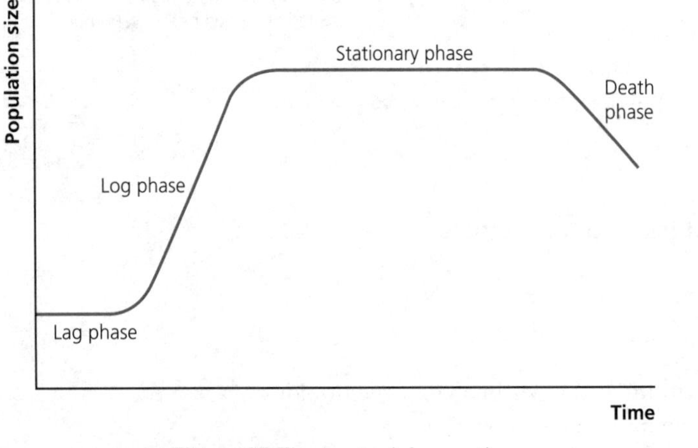

Figure 28 The bacterial growth curve

There are four phases of growth:

Lag phase During this phase the bacteria are taking in water and nutrients. They are also carrying out protein synthesis to produce enzymes to break down the nutrients present in the culture medium. The population is increasing slowly and the production rate is greater than the death rate.

Log (exponential) phase During this phase a lack of limiting factors allows the bacterial population to increase rapidly. The production rate is much greater than the death rate.

Stationary phase As the log phase ends the increase in population begins to slow and the curve levels out (plateaus), showing that the number of bacteria in the population remains constant. The culture is now in the stationary phase. Density-dependent limiting factors such as nutrient availability prevent the overall population size from increasing any further. Therefore the production rate is equal to the death rate.

Death phase Eventually the limiting factors become so great that the population begins to decline. During the death phase the production rate is less than the death rate.

In this investigation you will be able to observe the growth of a bacterial population and determine the exponential growth rate of the population.

Practical notes

A sample of bacteria is transferred to a liquid culture and the turbidity of the culture measured over time. The turbidity of the sample can be determined using a colorimeter. A more turbid solution has a greater absorbance of light than a less turbid solution, so the higher the absorbance the greater the number of bacteria in the sample. The absorbance can also be called the optical density (OD). The optical density is correlated to the number of bacteria, so a doubling of the optical density suggests a doubling of the number of bacteria present. You may also have the opportunity to measure the turbidity using a light sensor or colorimeter linked to a data logger.

This method of counting is known as an indirect total count. This is because the turbidity will be changed by both live and dead bacterial cells. A viable count method (such as dilution plating) will only count living bacteria.

Aseptic technique

Aseptic technique is important to ensure that the investigation is carried out safely and you obtain valid results. It is vital that this is carried out correctly to avoid contamination either to or from the environment and ensure that the culture remains pure (only has the bacteria you want in it).

- Sterilise the lab bench by wiping with a disinfectant.
- Wash your hands and/or wear gloves and a clean lab coat.
- Make sure that all apparatus and glassware are sterilised before use.
- Work near a Bunsen burner. You should light a Bunsen burner and keep it on the yellow flame near where you are working. This will create an updraft, which will reduce the risk of contamination by air-borne microbes. You will also need the Bunsen burner to flame-sterilise equipment.

- Reduce the amount of time any culture is exposed to the air by replacing lids as soon as possible.
- Flame the neck of the culture bottle — pass the culture bottle through the blue flame before taking a sample from it. Repeat after the sample has been taken and before the lid is replaced.
- Flame the neck of the culture vessel you are inoculating before and after inoculating.
- If using glass pipettes for the inoculation, dip the end into alcohol and then pass through the Bunsen burner. Be extremely careful not to leave the end of the pipette in the flame or it will melt.
- Make sure that culture media have been sterilised before inoculation with the microorganism.
- Dispose of all glassware by placing into a pressure cooker or autoclave; put all disposable items, such as syringes and pipette tips, into a container marked 'waste' to ensure that they are not reused. Plastic Petri dishes should be wrapped in a plastic bag, tied and sterilised in an autoclave.

Practical tip

Most of the apparatus you use should be sterilised by steam under pressure in a pressure cooker or autoclave at 121°C at 100 kPa for 15–20 minutes.

General points

Once you have inoculated the sterile medium the culture will need to be stirred. This can be done using a magnetic stirrer. The magnetic stirrer ensures an even distribution of the bacteria and culture medium. Settling out over time would lead to the absorbance changing as the top of the tube would become clearer while the bacteria concentrated at the bottom of the tube. This imbalance would make it difficult to accurately compare absorbances across the samples.

The magnetic stirrer also ensures that oxygen is distributed throughout the tube. This is important because the bacteria used are aerobic, and if any regions of the tube become oxygen depleted this would have a limiting effect on the growth of the bacterial population.

You should zero your colorimeter using a sample of the sterile culture media that has not been inoculated with any bacteria. This should be kept in the fridge between readings to ensure that no contaminating bacteria are able to grow in it.

Transfer a set volume of inoculated culture into a cuvette and measure the absorbance of a set wavelength. Ensure you are using the same volume of inoculated culture for each reading. Remember to follow aseptic techniques while taking each reading.

You should make at least five readings over 24 hours. Try to take several initial readings every 20–30 minutes. This will allow you to see the initial growth rate of the population. If your absorbance level rises above 2 then your readings will lack accuracy. If this occurs, dilute samples to ensure that the absorbance is below 2. You can then account for the dilution by multiplying the absorbance reading by the dilution factor.

From your results you should be able to:

- plot a graph of absorbance against time
- identify the lag and exponential phases

- estimate the time taken for the population to double during the exponential phase — this will be the generation time
- calculate the exponential growth rate using the formula:

$$k = \frac{\log_{10} OD_1 - \log_{10} OD_0}{\log_{10} 2 \times t}$$

where OD_0 is the absorbance at the start of the log phase, OD_1 is the absorbance reading after time t (use a reading towards the end of the log phase) and t is the time the culture has been growing.

You should also then plot your results on a second graph where absorbance is on a logarithmic scale. You can do this using either logarithmic graph paper or a computer program such as Excel.

Focus on maths skills

Logarithmic scales

A logarithmic scale is a non-linear scale that is useful for representing data when the range of values is very large (spanning several orders of magnitude). Logarithmic scales usually use logarithms to base 10 (also known as common logarithms), so that powers of 10 are marked along the vertical axis of the graph.

In A-level biology, logarithmic scales are mainly used for plotting microbial growth curves (Figure 29). Here are some important features of graphs of population growth plotted on a logarithmic scale:

- A straight line represents exponential growth (if the line is rising from left to right) or exponential decay (if the line is falling).
- A rising curve that is levelling off represents growth that is slower than exponential.
- A rising curve that bends upwards represents growth that is faster than exponential.

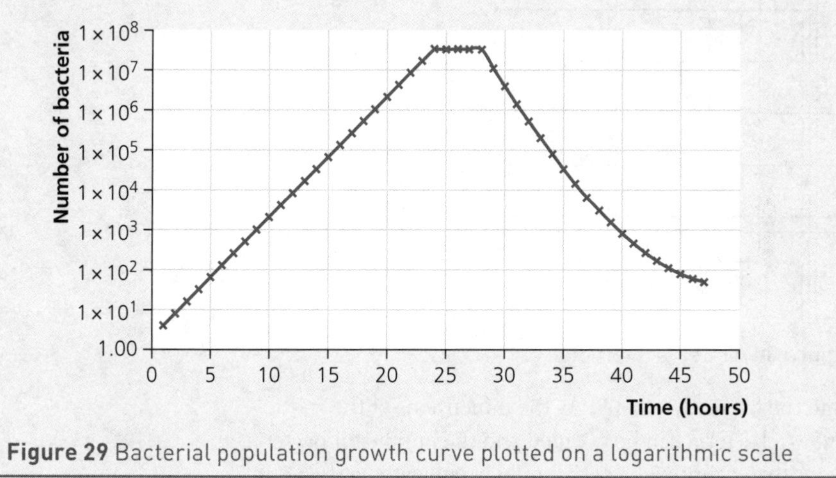

Figure 29 Bacterial population growth curve plotted on a logarithmic scale

Conclusion/evaluation

This investigation raises two main issues:

- Is the change in turbidity only due to an increase in the number of bacterial cells?
- What are the limitations of a total count?

For the first point, as long as the culture medium was sterile and free from other contaminants then we can be fairly sure that the change in the turbidity was due to the growth of the bacteria. However there is the possibility that the bacteria's metabolic processes, namely release of waste, could affect the amount of light absorbed by the solution. Overall this point should not have a huge bearing on the results.

The total count limitations are potentially more problematic. While we can be fairly sure that the increase in turbidity is due to an increase in the number of bacterial cells, we are unable to say if these cells are alive, dead or even just fragments of cells. This makes it difficult to make statements about when the population has reached the stationary or death phase, as in both these stages the turbidity and therefore the *OD* will continue to increase. This is because the total number of cells present in the culture will continue to increase (due to binary fission) even though the majority are likely to be dead. Therefore this method is most suitable for studying the lag/early log phase.

An improvement to this method could involve the use of a counting chamber such as a haemocytometer (Figure 30) or dilution plating to determine the number of bacterial cells present at a range of different *OD* readings. These readings could then be used to plot a calibration curve of bacterial numbers over *OD*, which would then allow the results of the investigation to be fitted to this curve to give an estimate of the number of cells at each time interval.

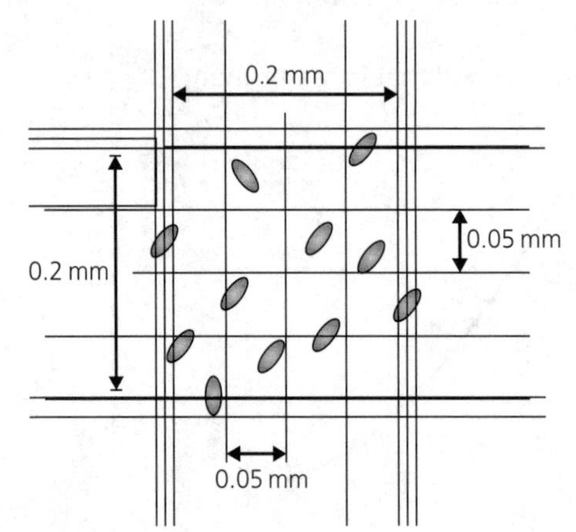

Figure 30 A haemocytometer

A sample of solution is placed onto the haemocytometer. As the dimensions of the haemocytometer chamber are known, bacteria can be counted and the number of bacteria per unit volume can be estimated. When counting bacteria in the haemocytometer, include any cells that touch or overlap the middle of the three lines at the top edge and right-hand side edge of the haemocytometer. Do not include any cells that touch or overlap the bottom edge or the left-hand edge. By sticking to these rules counting is consistent and repeatable. Following these rules means there are eight cells in this sample.

Assuming that the haemocytometer chamber has a depth of 0.1 mm, the total volume of the chamber is:

$$0.1 \times 0.2 \times 0.2 = 0.004\,\text{mm}^3$$

As there were eight cells in the count:

$$\frac{8}{0.004} = 2000$$

We can estimate that there are 2000 bacteria per mm^3.

The biggest improvement for the investigation would be to switch to a viable count method such as dilution plating. This involves serially diluting the sample and then transferring a known volume of each dilution to an agar plate. The plates are then incubated. Each bacterium forms a colony, which can then be counted on a suitably diluted plate to give an estimate of the number of bacteria in the original sample.

This method only counts living cells, so gives a greater degree of accuracy and also allows for the identification of the stationary and death phases. The disadvantages of this method are that it is more complex, requires a large amount of equipment (particularly agar plates) and is much more time-consuming. Turbidity is a useful, relatively simple and quick method of giving a good approximation of bacterial population growth, particularly during the lag and early log phase.

Core practical 13

Isolate an individual species from a mixed culture of bacteria using streak plating

CPAC 1a, 2a, 2b, 3a, 3b

In this practical you will be creating streak plates in order to isolate an individual species of bacterium. It is likely that you will be using *Micrococcus luteus*, which forms yellow colonies, and either *Escherichia coli* or *Bacillus megaterium*, which form white colonies.

Background

There is no particularly strong estimate of the number of bacterial species on Earth but it is likely to be millions, if not close to a billion. Bacteria show a wide range of shapes and sizes (morphology), but the incredible species richness is not reflected in bacterial morphology and many bacteria, and the colonies they form, look very similar.

There is enough difference in the colonies of some species for them to be identified and differentiated even with the naked eye. These differences include overall shape, shape of the colony margin, whether colonies are shiny or dull, and if the colony is flat or raised.

Streak plating is used to separate a single type of colony from a mixture. A sample of the colony is transferred onto a plate. The plate is then incubated. The bacteria that have been transferred will divide by binary fission to form a visible colony. These colonies can then be further sampled and additional streak plates created.

Practical notes

It is likely that your agar plates will be ready prepared. However, you should be aware of the process:

1 Agar powder is made to a 1% or 2% solution and boiled.
2 This is cooled until it is at a temperature of about 50°C. Nutrients may be added at this point unless a ready-made agar is used.
3 While still molten, at a temperature of 42–45°C, the agar is poured into Petri dishes (Figure 31).

You should label the base of agar plates you use in this investigation with your name and the date. Write this information in an arc close to the edge of the bottom of the plate.

Throughout this investigation you will need to use the aseptic techniques from core practical 12 along with some additional ones.

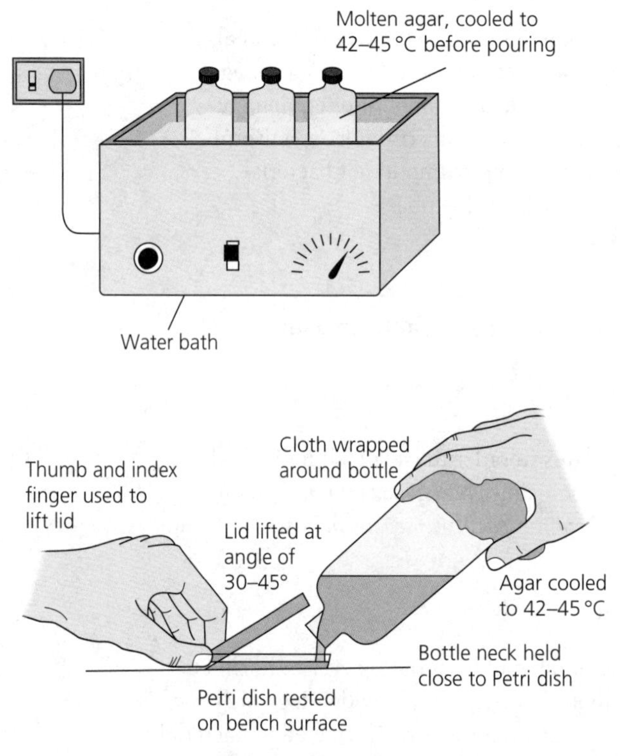

Figure 31 Preparing agar plates

Sterilising the inoculating loop

- The inoculating loop should be 'flamed' before being used to transfer a culture. Place the inoculating loop in the hottest part of the Bunsen burner flame until it glows red hot.
- Remove it from the flame and allow it to cool — this is important because if the red hot loop is placed on a culture it will kill the bacteria.
- You should sterilise the inoculating loop both before and after using it.

- When streaking only raise the Petri dish lid at a 45° angle (Figure 32). If you open the Petri dish fully there is a risk of the culture becoming contaminated with microbes in the air. By only opening the lid slightly this risk is reduced.
- When taping plates do not completely seal them. If a plate is completely sealed no air will be able to enter. The oxygen in the plate will quickly be used up by the respiring bacteria. This will cause the inside of the plate to become anaerobic and lead to the growth of anaerobic bacteria, which may be pathogenic.
- Do not put lids of tubes down onto benches. You should be able to unscrew the lid of the culture bottle using the thumb and little finger of your dominant hand and keep it there while using the inoculating loop.

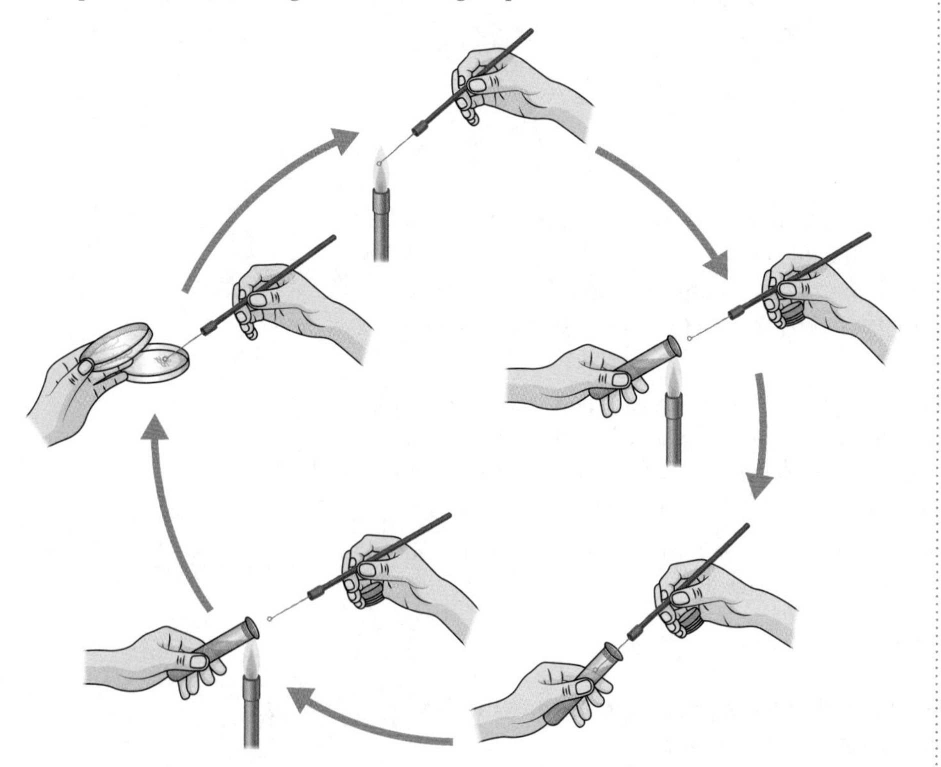

Figure 32 Practising the skills involved in transferring an inoculum from a broth culture to a Petri dish

When streaking the plate you should follow the pattern shown in Figure 33. When streaking, keep the loop at the same angle and turn the plate to allow you to move from A to B, B to C, C to D and D to E. It is important that each of the new streaks should pass through the previous streaks. The final streaks should not touch the first streaks. When streaking the plate do not press too heavily on the agar and 'gouge' it as this will lead to bacteria growing inside the agar as opposed to on its surface.

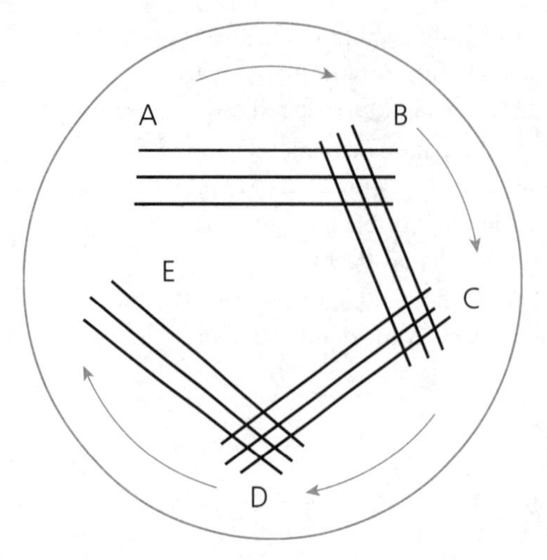

Figure 33 Agar plate streaking pattern

You should then incubate the plates. The plates should be incubated lid down. This prevents condensation forming on the lid. The plates should not be incubated above 30°C. Temperatures over this will provide optimum conditions for any pathogenic bacteria that may have contaminated the plate, leading them to divide and pose a potential hazard.

After the incubation period there should be a white and a yellow colony present. You should take a sample of yellow and streak it onto a plate labelled yellow, and take a sample of white and streak it onto a plate labelled white. To take a sample, dip a flamed and cooled inoculating loop into a colony of that colour on the final set of streaks.

Incubate these plates and then observe and sketch the colonies visible. Do not open the plates and ensure they are returned for sterilisation.

Core practical 14

Investigate the effect of gibberellin on the production of amylase in germinating cereals, using a starch agar assay

CPAC 1a, 2a–2d, 4a, 4b

Background

Gibberellins are plant hormones involved in germination of endospermic seeds such as cereal grains.

Cereal grains contain a starch store in the endosperm. At germination the starch is hydrolysed to a soluble form, so it can be transported to the growing points.

Gibberellins are released by the embryo. The gibberellins stimulate the production of hydrolytic enzymes such as amylase, which break down stored nutrients. Glucose and other nutrients that are products of this digestion diffuse to the embryo where they are used for aerobic respiration and growth.

Seeds that have had their embryos removed will still release amylase if they are soaked in gibberellins — this is what you will be doing in this investigation.

Practical notes

In this investigation you will be investigating the effect of different concentrations of gibberellin on the production of amylase by seeds. It is important to remove the embryos from the seeds (Figure 34). If the embryos were left on the seeds they would produce an unknown amount of gibberellins, making it impossible to control the independent variable.

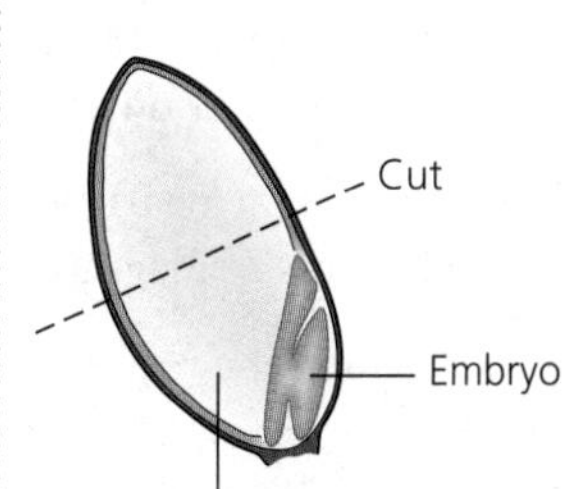

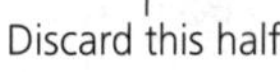

Figure 34 Removing the embryo from a seed

You will need to determine a range of values for the independent variable. The 'normal' concentration of gibberellin in germinating seeds is around 300 μg dm^{-3} (3×10^{-4} g dm^{-3}). One of your concentrations should be around this value along with at least four others. To give a good spread of results it may be appropriate to use serial dilutions to prepare a range of concentrations across several orders of magnitude.

Serial dilution

1. Use a graduated pipette or syringe to remove 1 cm^3 of the stock gibberellin solution and put into a test tube.
2. Add 9 cm^3 of distilled water and stir or invert the test tube. Assuming that the stock solution was 1 g dm^{-3}, this is now a 1×10^{-1} g dm^{-3} solution.
3. Repeat this process to give five different concentrations of gibberellin.

An example of a possible range of concentrations to use is shown in Table 10.

Table 10 Gibberellin concentrations

Concentration of gibberellin/g dm^{-3}
1×10^{-1}
1×10^{-2}
1×10^{-3}
1×10^{-4}
1×10^{-5}

The production of amylase is assessed using a starch agar assay. Starch causes iodine to change colour to blue/black. Starch is added to the agar before the plate is prepared so permeates the entire plate.

If the cereal grains produce amylase it will diffuse into the agar. The amylase will then break the starch down into maltose.

The plate is then flooded with iodine. Areas where starch remains in the agar will turn blue black. Areas where the starch has been broken down will be 'clear'. By comparing the areas of the clear regions produced by the different concentrations of gibberellin, the relative amount of amylase released can be approximated.

The area of the clear regions will be the dependent variable, with larger areas indicating an increased release of amylase by the seeds. The clear regions should be roughly circles, so their areas can be calculated using the formula for the area of a circle.

Practical tip

You can make up serial dilutions using other factors, for example by a factor of 2 by transferring 5 cm^3 each time to make a maximum volume of 10 cm^3.

Focus on maths skills

Area of a circle

area of a circle = πr^2

So, if the clear area has a diameter of 25 mm:

$$\text{radius} = \frac{\text{diameter}}{2} = \frac{25}{2} = 12.5\,\text{mm}$$

$$\text{area} = \pi r^2 = \pi \times 12.5^2 = 491\,\text{mm}^2$$

Sterile techniques still need to be used during this practical as microorganisms could contaminate the agar. This can include dipping the seeds into 3% sodium hypochlorite solution for 5 minutes before starting the investigation.

You could use a logarithmic scale for the concentration of gibberellin as it runs over several orders of magnitude. For more detail on logarithmic scales, see page 57.

You can pool the results of the class and perform a Spearman's rank correlation coefficient test. Even if other members of the class use different concentrations, the Spearman's rank correlation coefficient test can be used.

Focus on maths skills

Spearman's rank correlation coefficient

This test is based on one of the most common correlation coefficients used in biology. It provides a way of measuring the strength of a relationship between paired data. In this investigation the paired data comprise the concentration of gibberellin and the area of the clear region that it produced. See, for example, Table 11.

Table 11 Class data

Concentration of gibberellin/g dm^{-3}	Area of clear region/mm^2
2×10^{-4}	131
2×10^{-1}	262
4×10^{-4}	101
1×10^{-4}	119
1×10^{-2}	275
5×10^{-3}	205
1×10^{-6}	0
6×10^{-3}	216
1×10^{-3}	189
5×10^{-2}	110

Step 1: The first step when carrying out a statistical test is to write down the null and alternative hypotheses.

- Null hypothesis: there is no significant correlation between concentration of gibberellin and the area of the clear region of agar.
- Alternative hypothesis: there is a significant correlation between concentration of gibberellin and the area of the clear region of agar.

Step 2: Rank order the values of the independent variable (concentration of gibberellin) making sure that the data values remain linked to the corresponding values of the dependent variable.

Make a table for this, and create a column showing the rank order (R_1) of each value of the independent variable (Table 12).

Table 12 Class data ranked by concentration of gibberellin

Concentration of gibberellin/g dm^{-3}	R_1	Area of clear region/mm^2
1×10^{-6}	1	0
1×10^{-4}	2	119
2×10^{-4}	3	131
4×10^{-4}	4	101
1×10^{-3}	5	189
5×10^{-3}	6	205
6×10^{-3}	7	216
1×10^{-2}	8	275
5×10^{-2}	9	110
2×10^{-1}	10	262

Step 3: Now rank order the values of the dependent variable (clear area). Create a new column showing the rank order (R_2) of each value of the dependent variable (Table 13).

Table 13 Class data with R_1 and R_2

Concentration of gibberellin/g dm^{-3}	R_1	Area of clear region/mm^2	R_2
1×10^{-6}	1	0	1
1×10^{-4}	2	119	4
2×10^{-4}	3	131	5
4×10^{-4}	4	101	2
1×10^{-3}	5	189	6
5×10^{-3}	6	205	7
6×10^{-3}	7	216	8
1×10^{-2}	8	275	10
5×10^{-2}	9	110	3
2×10^{-1}	10	262	9

Step 4: For each row, calculate the difference d between the ranks ($d = R_1 - R_2$) and enter the results in a fifth column. A good way of checking if you have calculated the differences correctly is to sum the values in the d column. The sum should be zero, and if it is not then you should go back and check your working.

Then, square each d and enter the result (d^2) in a sixth column (Table 14).

Table 14 Class data showing difference in ranks squared

Concentration of gibberellin/g dm^{-3}	R_1	Area of clear region/mm^2	R_2	Difference in ranks, $d = R_1 - R_2$	d^2
1×10^{-6}	1	0	1	0	0
1×10^{-4}	2	119	4	-2	4
2×10^{-4}	3	131	5	-2	4
4×10^{-4}	4	101	2	2	4
1×10^{-3}	5	189	6	-1	1
5×10^{-3}	6	205	7	-1	1
6×10^{-3}	7	216	8	-1	1
1×10^{-2}	8	275	10	-2	4
5×10^{-2}	9	110	3	6	36
2×10^{-1}	10	262	9	1	1

Step 5: Calculate Spearman's rank correlation coefficient, r_s:

$$r_s = 1 - \frac{6\sum d^2}{n(n^2 - 1)}$$

where n is the number of pairs of data values. Sum the values in the d^2 column to get $\sum d^2 = 56$. Then substitute this into the formula:

$$r_s = 1 - \frac{6 \times 56}{10(10^2 - 1)}$$

$$= 1 - \frac{336}{990} = 0.661$$

Step 6: Interpret this value. Spearman's rank correlation coefficient can tell us two things:

- Direction of correlation — if the value is positive, there is a positive correlation between the two variables. If the value is negative, there is a negative correlation between the two variables. As 0.661 is a positive value, there is a positive correlation between average concentration of gibberellin and the area of the clear region.
- Strength of correlation — by comparing r_s with the critical value from a table, we can test whether the correlation is significant or not.

There are two types of test, one-tailed and two-tailed. If the direction of the correlation is stated in the alternative hypothesis (i.e. we take it as known), then we use a one-tailed test. If the direction of the correlation is not stated in the alternative hypothesis (i.e. we do not know it beforehand), then we use a two-tailed test. As we do not state a direction in our alternative hypothesis we should use a two-tailed test.

Table 15 is an extract of critical values for a two-tailed Spearman's rank correlation coefficient test.

- If r_s is lower than the critical value, then accept the null hypothesis and reject the alternative hypothesis.
- If r_s is higher than the critical value, then reject the null hypothesis and accept the alternative hypothesis.

Table 15 Critical values for a two-tailed Spearman's rank correlation coefficient test

Number of pairs of observations	Significance level, p	
	0.1	0.05
8	0.643	0.783
9	0.600	0.700
10	0.564	0.648

In this example, the number of pairs of observations is 10, so at the 0.05 significance level typically used in biology, the critical value is 0.648.

Because $r_s = 0.661 > 0.648$, we can conclude that in this instance there is a significant positive correlation between the concentration of gibberellin and the area of the clear region.

Core practical 15

Investigate the effect of different sampling methods on estimates of the size of a population

CPAC 1a, 2a–2d, 4a, 4b, 5a

In this investigation you will be comparing different sampling techniques involving quadrats.

Background

Quadrats are frames used to sample non-motile organisms such as plants, fungi and shore animals such as barnacles and limpets. The two types of quadrat used in this investigation are:

- **frame quadrat** — square of a known area. These are placed on the ground (either randomly or systematically) and then used to estimate population density or percentage cover. Common sizes of quadrats used in A-level investigations are $0.25\,m^2$ and $1\,m^2$.
- **point quadrat** — a frame with holes through which long needles are lowered to the ground below (Figure 35). They can be used to estimate percentage cover by finding the proportion of needle points that touch (hit) the species being studied. If point quadrats are unavailable you can use a frame quadrat that has been subdivided into smaller squares. If using a frame quadrat the 'points' are the corners of the small squares (where the wires overlap).

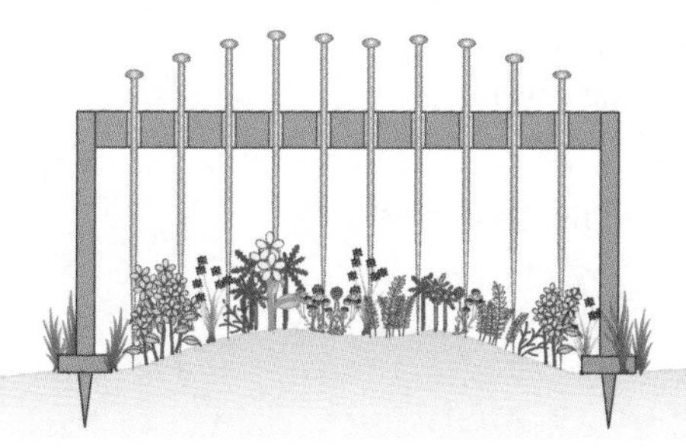

The frame is randomly placed a large number of times, the 10 pins are lowered in turn on to vegetation and the species (or bare ground) recorded.

Figure 35 A point quadrat

Quadrats can be used in random or systematic sampling. In this investigation you will be using random sampling. This does not mean throwing the quadrat and seeing where it lands! You will use a random number generator to determine coordinates for the quadrat within a larger sample area.

Practical notes

Random sampling

Use two tape measures to measure out a suitable sample area (e.g. 10 m by 10 m). You will need to place your quadrat randomly within this area. Imagine that the area has been divided into 1 m by 1 m squares.

Use a random number generator to give two numbers between 1 and 10. Use these numbers as coordinates on the grid, for example the numbers 4 and 6 (Figure 36). Place the quadrat at the corner of these points.

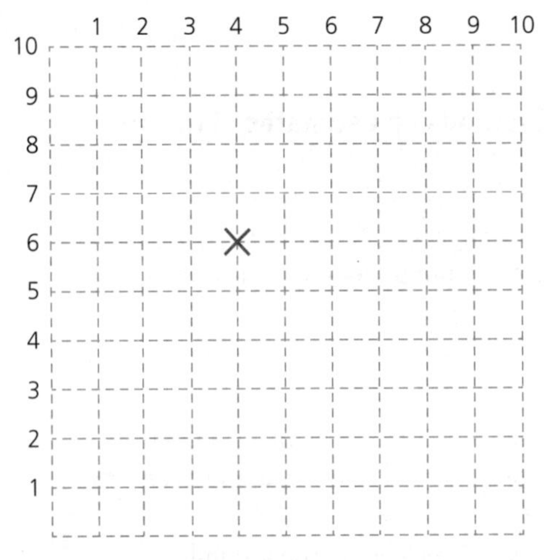

Figure 36 Sample area with coordinates

You will be studying the population of a single species, so you first need to select a species to study. This should be one that is found within your sample area and is easy for you to identify. You should pick a species that has at least 5% cover (so a species that is relatively common within the sample area). Some good examples include dandelions, ribwort plantain, greater plantain and catsear.

When you have placed your quadrat you should record the **percentage cover** — this is the percentage of the quadrat area that is occupied by individuals of a particular species. This measure is particularly useful for very numerous species such as grasses. For example, each of the 0.25 m^2 quadrats was divided into 25 equal squares. In quadrat 1, grass filled 21 of the smaller squares. What was the percentage cover of grass in this quadrat?

To find the percentage cover, take the area of the quadrat that is covered by the organism and divide by the total area of the quadrat:

$$\frac{21}{25} \times 100\% = 0.84 \times 100\% = 84\%$$

So the percentage cover of grass in this quadrat is 84%.

If using the point quadrat (or a subdivided frame quadrat) to estimate the percentage cover, place the quadrat down and record how many times a point (or grid intersection) touches the plant you are studying — a 'hit'. If a point touches a plant as you lower it then count that as a hit, but don't record more than one hit of the same species by a single point: for example, if the same species is touched three times by the same point as you lower it that still counts as one 'hit'.

You can also record **species density** — the number of individuals of a certain species per unit area. To do this, count the number of individuals of your chosen plant that are in the quadrat. When counting the plants trace down to the stem to ensure that you are counting each individual plant rather than the same plant with multiple stems. If you count every plant that crosses all the edges of the quadrat it may give you an inaccurate count. To avoid this do not count plants that cross the right-hand and lower sides of the quadrat — only count the plants that cross the top and left-hand sides.

You should now replace the 0.25 m^2 quadrat with a 1 m^2 quadrat and again measure the species density.

Repeat the above at least 10 times within your sample area so that you have at least 10 percentage cover estimates using each frame and point method and 10 species density measurements using the 0.25 m^2 quadrat and 10 using the 1 m^2 quadrat.

You should then tabulate these data (Table 16).

Table 16 Example data for a random sampling study

Quadrat area/m^2		Sample									
		1	2	3	4	5	6	7	8	9	10
0.25 (with 16 points)	Point 'hits'	3	0	0	1	1	5	0	0	2	4
0.25	Percentage cover (frame quadrat)	12	4	0	4	12	20	4	0	8	20
0.25	Number of plants	8	1	0	1	2	7	1	0	3	6
1.00	Number of plants	33	1	4	5	10	39	4	0	7	27

From these data you can then calculate the percentage cover from the point quadrat and compare this with the estimates given by the frame quadrat. To calculate the percentage cover from the point quadrat multiply the 'hits' value by:

$$\frac{100}{\text{number of points on the quadrat}}$$

In this example, this means multiply the 'hits' value by 100/16 = 6 (1 s.f.). See Table 17.

Table 17

Type of quadrat	Percentage cover									
Point quadrat	18	0	0	6	6	30	0	0	12	24
Frame quadrat	12	4	0	4	12	20	4	0	8	20

You can also compare the species density estimates from the 0.25 m^2 quadrat and the 1 m^2 quadrat. To do this multiply the values from the 0.25 m^2 quadrat by 4 (Table 18).

Table 18

Quadrat area/m^2	Density/plants m^{-2}									
0.25	32	4	0	4	8	28	4	0	12	24
1.00	33	1	4	5	10	39	4	0	7	27

You should now compare the results from each method. A good way of doing this is using a bar chart with range bars on. To see if there is a significant difference between the sampling methods you should use a statistical test such as the Student's *t*-test.

Focus on maths skills

The Student's *t*-test

The *t*-test can be used to determine whether there is a significant difference between two sets of data by comparing their means.

Step 1: As in all statistical tests, start by writing a null hypothesis and an alternative hypothesis:

- Null hypothesis: There is no significant difference between the percentage covers found using the frame quadrat and the point quadrat.
- Alternative hypothesis: there is a significant difference between the percentage covers found using the frame quadrat and the point quadrat.

Step 2: To apply the *t*-test, we need the mean $(\bar{x})$ and variance (s^2) of each data set:

$$\text{mean of point quadrat} = (18 + 0 + 0 + 6 + 6 + 30 + 0 + 0 + 12 + 24)/10 = 9.6$$

$$\text{mean of frame quadrat} = (12 + 4 + 0 + 4 + 12 + 20 + 4 + 0 + 8 + 20)/10 = 8.4$$

The variance is the sum of the squared deviations from the mean, divided by the number of data points. The best way of calculating this is using a table. Table 19 shows this for the point quadrat data.

Table 19

Percentage cover	18	0	0	6	6	30	0	0	12	24
Deviation from mean $(x - \bar{x})$	8.4	-9.6	-9.6	-3.6	-3.6	20.4	-9.6	-9.6	2.4	14.4
$(x - \bar{x})^2$	70.56	92.16	92.16	12.96	12.96	416.16	92.16	92.16	5.76	207.36

A good way of checking if the deviations from the mean are correct is to sum them. This sum should always equal zero. Now add together the squared deviations from the mean (the values in the last row of the table above) and divide by the number of data points to obtain the variance.

$$\text{variance of point quadrat} = \frac{70.56 + 92.16 + 92.16 + 12.96 + 12.96 + 416.16 + 92.16 + 92.16 + 5.76 + 207.36}{10}$$

$$= 109.44$$

Table 20 shows the data for the frame quadrat.

Table 20

Percentage cover	12	4	0	4	12	20	4	0	8	20
Deviation from mean $(x - \bar{x})$	3.6	−4.4	−8.4	−4.4	3.6	11.6	−4.4	−8.4	−0.4	11.6
$(x - \bar{x})^2$	12.96	19.36	70.56	19.36	12.96	134.56	19.36	70.56	0.16	134.56

$$\text{variance of frame quadrat} = \frac{12.96 + 19.36 + 70.56 + 19.36 + 12.96 + 134.56 + 19.36 + 70.56 + 0.16 + 134.56}{10}$$

$$= 49.4$$

Step 3: Use the *t*-test formula:

$$t = \frac{\bar{x}_1 - \bar{x}_2}{\sqrt{\frac{s_1^2}{n_1} + \frac{s_2^2}{n_2}}}$$

where:

$\bar{x}_1$ = higher mean

$\bar{x}_2$ = lower mean

s_1^2 = variance of data set with the higher mean

s_2^2 = variance of data set with the lower mean

n_1 = number of values in data set with the higher mean

n_2 = number of values in data set with the lower mean

Substituting the values found in step 2:

$$t = \frac{9.6 - 8.4}{\sqrt{\frac{109.44}{10} + \frac{49.4}{10}}}$$

$$= \frac{1.2}{\sqrt{10.944 + 4.94}}$$

$$= \frac{1.2}{\sqrt{15.884}}$$

$$= 0.301$$

Step 4: We now need to compare this value with the critical value from a table. First we need to determine the degrees of freedom. For *t*-tests this is obtained by subtracting 1 from the number of data points in each set and adding them together.

In this example, each data set has 10 values, so the degrees of freedom value is (10 − 1) + (10 − 1) = 18.

Step 5: Read off the critical value from Table 21. In biology we normally use the 0.05 level of significance.

Table 21 Critical values for the Student's *t*-test

Degrees of freedom	Significance level *p*					
	0.1	0.05	0.02	0.01	0.002	0.001
14	1.761	2.145	2.624	2.977	3.787	4.140
15	1.753	2.131	2.602	2.947	3.733	4.073
16	1.746	2.120	2.583	2.921	3.686	4.015
17	1.740	2.110	2.567	2.898	3.646	3.965
18	1.734	2.101	2.552	2.878	3.610	3.922
19	1.729	2.093	2.539	2.861	3.579	3.883

Looking at the row for 18 degrees of freedom and the column for significance level 0.05 gives a critical value of 2.101.

Step 6: Compare the calculated *t* value with the critical value. If the calculated *t* value is *lower* than the critical value, then *accept the null hypothesis* and *reject the alternative hypothesis*. If the calculated *t* value is *higher* than the critical value, then *reject the null hypothesis* and *accept the alternative hypothesis*.

In this case $0.301 < 2.101$. So accept the null hypothesis and reject the alternative hypothesis. You can conclude that there is no significant difference between the percentage covers found using the frame quadrat and the point quadrat.

Extension: Simpson's diversity index

By pooling class data it is possible to calculate the Simpson's diversity index for the area. For example, Table 22 shows the results of an investigation into the species frequency of some common plants on a school field.

Table 22

Species	Number (*n*)
Taraxacum sp. (dandelion)	2
Bellis perennis (daisy)	3
Grass	47

The equation for Simpson's diversity index *D* is:

$$D = \frac{N(N-1)}{\Sigma n(n-1)}$$

where N = total number of organisms of all species found and n = number of individuals of each species. Σ means 'sum' or 'add together' values for the different species found in that area.

$$N = 2 + 3 + 47$$
$$= 52$$

It is best to calculate this using a table to find $n - 1$ and then $n(n - 1)$ — see Table 23.

Table 23

Species	Number (n)	$n - 1$	$n(n - 1)$
Taraxacum sp. (dandelion)	2	1	2
Bellis perennis (daisy)	3	2	6
Grass	47	46	2162

$D = 52(52 - 1)/(2 + 6 + 2162)$

$D = 2652/2170$

$D = 1.22$

Core practical 16

Investigate the effect of one abiotic factor on the distribution or morphology of one species

CPAC 1a, 2a–2d, 3a–3c, 4a, 4b, 5a, 5b (dependent on investigation)

Practical notes

In this investigation you will be studying the effect of an abiotic factor on the distribution or morphology (form and structure) of an organism. You will need to plan this investigation yourself and it will be dependent on what is available to you. Some possibilities are outlined below.

The effect of light intensity on the distribution of a plant species

In this investigation you can use a belt transect as opposed to the random sampling method carried out in core practical 15. You will be investigating the effect of an environmental gradient on the distribution of an organism. A transect sample is taken systematically in a line:

1 First pick a site that shows an obvious gradient of light intensity, for example a wood with a shaded area and a more open area.

2 Lay out a tape measure (e.g. 20 m) in a straight line along the gradient you have identified. This will be your transect.

3 Identify a species that seems to vary in abundance along the transect.

4 Use a light meter to measure the light intensity at ground level at regular intervals along the transect.

5 Place quadrats at the same intervals along the transect and use them to determine either the species density or percentage cover.

The effect of light intensity on the morphology of plants

Plant growth is affected by light intensity because light intensity affects the rate of photosynthesis. By comparing plants in direct sunlight and more shady areas you will be able to see if this difference in light intensity has affected the morphology of the plants. In this case light intensity is the independent variable.

Measuring leaf size is one of the most common ways of studying the effect of an abiotic factor on plant morphology. Size itself is not a variable, however — there are different measurements of size, including:

- width
- length
- area
- dry mass

Other possible dependent variables include:

- stem internode length — a node is a point on the stem at which leaves are attached. The internode length is the distance between the nodes. A faster-growing plant has a longer internode length than a slower-growing one.
- petiole length — the petiole is the stalk that joins the leaf to the stem
- number of serrations on leaf edges — count the number of serrations around the edges of the leaf

It may also be more appropriate to consider relative dimensions as an independent variable, for example leaf length-to-width ratios.

The effect of shore height or wave exposure on the morphology of rocky shore organisms

Many rocky shore organisms such as limpets, barnacles, dog whelks and seaweed vary in morphology depending on their position on the shore. The main factors that influence this are the tidal height (which affects how long they are immersed in the water for and at what depth) and the exposure to waves and wind (and the forces the organisms experience due to these). These different factors could be used as independent variables in an investigation.

The dependent variables in this investigation will vary with the organisms. Some examples of possible areas of investigation include shell heights, surface-area-to-volume ratios of organisms and number of air bladders on some species of seaweed, such as bladderwrack, *Fucus vesiculosus*.

Mineral ion availability and morphology of seedling growth

A lab-based alternative to the fieldwork in this investigation is to study the effect of a lack of mineral ions on plant growth. Some examples of minerals and the effects on plants when they are lacking include:

- **magnesium** — an essential component of chlorophyll. Plants that are deficient in magnesium develop chlorosis (yellowing of leaves).
- **nitrate** — plants require nitrate to produce amino acids, nucleotides and chlorophyll. A lack of nitrate leads to reduced growth and chlorosis.
- **phosphate** — plants require phosphate to produce phospholipids. A lack of phosphate leads to reduced growth.

Seeds such as mung beans can be suspended above solutions deficient in these minerals and the growth of the seedlings observed over time. In this case the independent variable would be ions the plants were deficient in. As with the investigation into the effect of light intensity, there are a number of different possible dependent variables including leaf area, mass, root length or stem length. Relative dimensions could also be studied, such as the ratio of stem length to root length.

General points

This investigation should bring together many of the skills you have developed throughout the course. Steps you should take in this investigation include the following:

- Do research on your chosen topic, focusing on high-quality sources, including articles in scientific journals if possible.
- Use your research to write a hypothesis and null hypothesis, along with an explanation of how you came up with them.
- Write a detailed, step-by-step plan of your investigation. Try to test the methods you will use in a small pilot study. Use this to inform your planning. You should also reference your initial research throughout your plan. Your plan should include:
  - the independent variable and the range you will be using. How will you determine different levels of light intensity or wave exposure, for example?
  - what your dependent variable is and how it will be measured, including the units. This section should also consider how you will avoid bias when collecting your data.
  - control variables. You should state how you will control these and, if they cannot be controlled, how you will measure them and take them into account.
  - an equipment list
  - a risk assessment
  - your choice of statistical test and justification for this choice
- Carry out your investigation and record the results in an appropriate table.
- Draw a suitable graph of your results.
- Carry out the appropriate statistical test.
- Write a conclusion. This should explain your results using your biological knowledge. Your conclusion should be supported by scientific references.
- Evaluate your investigation. This will involve identifying any anomalous results and explaining why you think they occurred. You should comment on the precision and accuracy of your results, and state values for the uncertainty of key measurements.
- You should reference all research in the correct way (see page 29 for more details on this).

Questions & Answers

Exam format

At AS there are two examination papers, each worth 50% of the total AS grade. Paper 1 covers Core Cellular Biology and Microbiology and paper 2 covers Core Physiology and Ecology. At A-level there are three exam papers: paper 1: Advanced Biochemistry, Microbiology and Genetics, paper 2: Advanced Physiology, Evolution and Ecology and paper 3: General and Practical Principles in Biology. Paper 1 and paper 2 are worth 30% each and paper 3 is worth 40%. Paper 3 will include synoptic questions, which may cover two or more different topics. The paper will also include questions that assess your conceptual and theoretical understanding of experimental methods.

About this section

This section contains questions on some of the core practicals. These may appear in papers 1 and 2 in the AS exams and in paper 3 in the A-level exams. They are written in the same style as the questions in the exam so they will give you an indication of what you can expect. After each question there are answers by two different students.

Comments on the questions are preceded by the icon ⓔ. They offer tips on what you need to do to gain full marks. All student responses are followed by comments, indicated by the icon ⓔ, which highlight where credit is due. In the weaker answers, they also point out areas for improvement, specific problems and common errors such as lack of clarity, irrelevance, misinterpretation of the question and mistaken meanings of terms.

Question 1 Plant cell water potentials

(a) The graph below shows the change in mass of potato cylinders in sucrose solutions with different osmotic potentials.

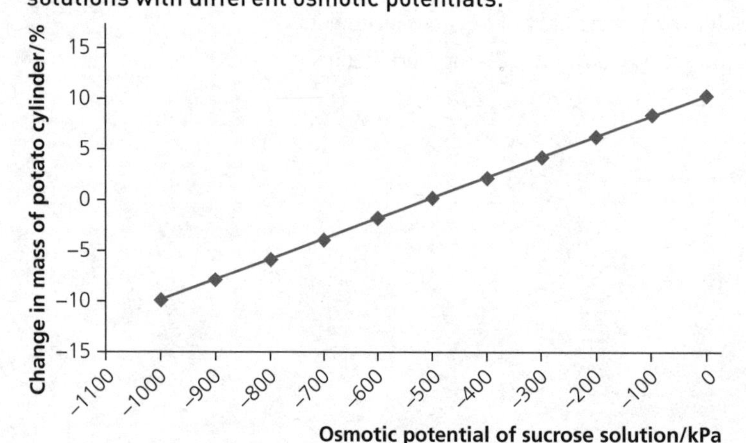

(i) Explain the importance of using percentage change in mass in this investigation as opposed to change of mass in grams. (1 mark)

(ii) The experiment was repeated and the results in the table below were obtained. Calculate the percentage change in mass for sample D and complete the table. (1 mark)

Sample	Initial mass/g	End mass/g	% change in mass
A	1.73	1.62	−6.36
B	2.07	1.87	−9.66
C	1.85	1.75	−5.41
D	2.11	2.20	
E	2.01	2.17	7.96

(iii) Using the graph estimate the osmotic potentials that produced these changes in mass. (2 marks)

The diagram below shows the water potential of three plant cells.

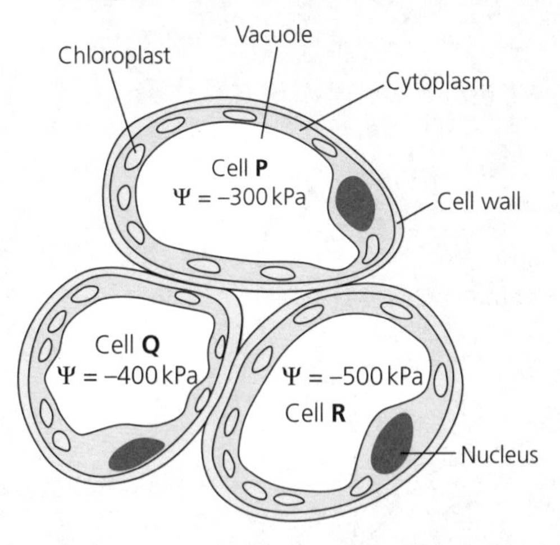

(b) Draw arrows to show the movement of water on the diagram above. (2 marks)

The photos below shows plant cells that have been put in two different concentrations of sucrose solution, P and Q.

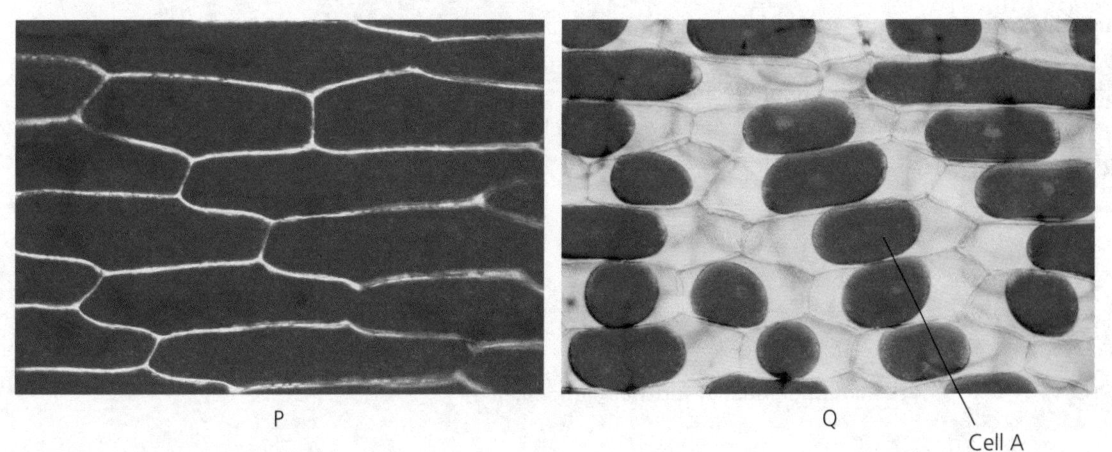

P Q

(c) (i) Which of the solutions had the highest concentration of sucrose? Explain your answer. (2 marks)

(ii) The osmotic potential of the cells in solution Q is –200 kPa. What is the water potential of cell A? Explain how you arrived at your answer. (4 marks)

ⓔ It is easy to make mistakes in water potential questions. Do not rush into an answer. Go through each part of the question methodically and always check back on the work you have done. Question (c) (ii) is particularly challenging. If you get to a question like this remember that the answer must be based on something in the specification, so take a bit of time to think around the question and see if you can work out how it relates to what you know.

Student A

(a) (i) Not all the potato samples had the same starting mass. Therefore in order to compare the changes in mass the % change in mass must be calculated.

ⓔ **1/1 mark awarded** This is the correct answer and gains the 1 mark available.

(ii) 95

ⓔ **0/1 mark awarded** Student A has incorrectly calculated the percentage change in mass for sample B. This should have been obvious as the value is so much greater than all the rest.

(iii)

Sample	Osmotic potential/kPa
A	–800
B	0
C	–750
D	Greater than 0
E	–100

ⓔ **1/2 marks awarded** The error in part (ii) has then led student A to give an incorrect value for the osmotic potential for D. Again this should have been obvious as such a change would be caused by an osmotic potential much greater than 0. Student A therefore scores 1 of the 2 marks available for determining the osmotic potentials.

(b)

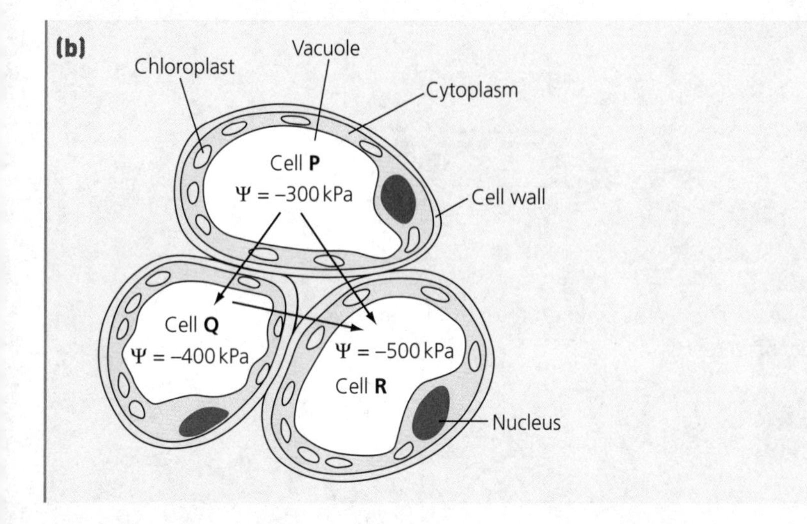

2/2 marks awarded The direction of movement of the water is correctly shown, going from a higher (less negative) water potential to a lower (more negative) water potential.

(c) (i) Solution P has the highest sucrose concentration of the two solutions. You can tell from the appearance of the cells.

1/2 marks awarded Student A has correctly identified solution P as having the highest concentration of sucrose. However, they have failed to explain why.

(ii) The water potential of the cell is 0 kPa because the cells are plasmolysed.

1/4 marks awarded Student A clearly does not know how to answer this question but scores a mark by using the keyword 'plasmolysed' in the correct context. This shows how it is possible to pick up marks even in complex questions by using appropriate keywords, as long as they are in the correct context.

Student B

(a) (i) The potato samples did not all start with the same mass. Therefore to make a fair comparison of the changes in mass the % change in mass must be calculated for each sample.

1/1 mark awarded A clear and correct answer.

(ii) 4.27

1/1 mark awarded Student B has correctly calculated the percentage change in mass.

(iii)

Sample	Solute potential/kPa
A	-800
B	0
C	-750
D	-300
E	-100

e **2/2 marks awarded** All the answers are correct. When asked to estimate in a question there will often be a margin of error given.

(b)

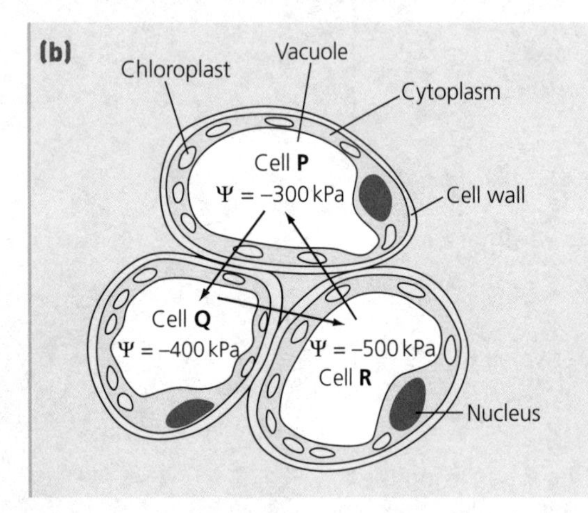

e **1/2 marks awarded** One of the arrows is pointing in the wrong direction, so student B loses one of the available marks. This incorrect answer to a simple question by an able student highlights the importance of checking answers involving water potential. The use of negative numbers can make it easy to make errors.

(c) (i) Solution Q has the highest sucrose concentration and therefore the lowest water potential of the two solutions as the cells in solution Q are plasmolysed, therefore water must have left the cell by osmosis, from a high to a low water potential.

e **2/2 marks awarded** This is a detailed answer that scores both available marks. It would also be possible to score marks by explaining why solution P does not have the highest concentration of sucrose — the water potential of this solution must be high as water moves into the cell so it is turgid.

(ii) The water potential of the cell is –200 kPa. This is the same as the osmotic potential. As the cell is plasmolysed the turgor pressure is zero. This is because the cytoplasm is not pushing against the cell wall.

e **4/4 marks awarded** Student B has realised that as the cell is plasmolysed there is no turgor pressure, so the osmotic potential is equal to the water potential.

Question 2 Plasma membrane permeability

Beetroot cells contain the red pigment betalain. The betalain is unable to pass through the cell's plasma membrane. Samples of beetroot were placed in distilled water and heated to different temperatures. The beetroot was removed from the samples and then the transmission of green light was measured in a colorimeter. The results are shown in the table below.

Temperature/°C	Transmission of light/% 1	2	3	4	5	Mean transmission/%	Standard deviation
20	98.6	99.0	100.0	96.0	96.0		
30	87.5	89.8	79.0	85.6	89.2	86.22	3.89
40	75.4	67.0	78.2	76.5	33.0	66.02	16.96
50	46.4	53.5	49.3	45.0	47.4	48.32	2.94
60	18.4	26.0	22.4	21.0	23.2	22.20	2.50

(a) **Calculate the mean and standard deviation for 20°C.** (2 marks)

(b) **Explain the standard deviation value for 40°C.** (2 marks)

(c) **Describe the trend of the mean results.** (1 mark)

(d) **Explain the trend you described in part (c).** (3 marks)

(e) **A similar pattern of results could be obtained using increasing concentrations of an organic solvent such as acetone. Explain why.** (2 marks)

e This question requires you to calculate and interpret standard deviation. Make sure you can do this as it is one of the mathematical skills required in the A-level biology specification.

Student A

(a)

Temperature/°C	Mean transmission/%	Standard deviation
20	97.92	1.87

e **1/2 marks awarded** The 20°C standard deviation value has been incorrectly calculated and so the student loses a mark. This shows the importance of checking all your calculations carefully. The mean value is correct so the student scores 1 mark.

(b) The standard deviation is greater than the others because reading number 5 is much lower than the others so is probably anomalous.

e **2/2 marks awarded** Student A has identified the standard deviation as much greater than the other values and that reading number 5 is the cause of this, so scores both available marks.

(c) As the temperature increases the mean transmission also decreases.

e **1/1 mark awarded** This answer is correct and scores a mark.

(d) As the temperature increases the membrane becomes more fluid so the transmission decreases.

e **1/3 marks awarded** This is a simple answer so only scores 1 mark. It needed further development, as in student B's answer, to get the other 2 marks.

(e) Phospholipids are highly soluble in organic solvents so the phospholipids dissolve when the solvents are added. This means that the betalain passes out of the cell, decreasing the transmission.

e **2/2 marks awarded** This answer scores both marks as student A explains the effect of the phospholipids being soluble in the organic solvent.

Student B

(a)

Temperature/°C	Mean transmission/%	Standard deviation
20	97.92	1.63

e **2/2 marks awarded** Both values are correct.

(b) The standard deviation at 40°C is much greater than the others as the transmission results at the temperature have a larger range.

e **1/2 marks awarded** This is correct as far as it goes, but student B misses out on the second mark because of a lack of detail. Rather than just stating that the range is larger, the answer also needs to say why — reading 5 is much lower than the others.

(c) As the temperature increases the mean transmission decreases.

e **1/1 mark awarded** The trend is straightforward in these data, so this simple answer scores the available mark. If the data showed a clear drop then appeared to plateau it would be important to describe each of these stages.

(d) As the temperature increases the membrane becomes more fluid so the betalain is able to move out of the plasma membranes of the beetroot cells. This decreases the transmission as the betalain absorbs the light. At higher temperatures the proteins in the membrane denature, leaving gaps in the membrane, so more betalain leaves, further decreasing the transmission.

e **3/3 marks awarded** This is a good answer, which achieves full marks.

(e) Phospholipids are soluble in organic solvents.

e **1/2 marks awarded** This answer scores only 1 mark. It is not developed enough to get the second mark.

Question 3 Enzyme-controlled reactions

The graph below shows the mass of product formed by an enzyme-controlled reaction over time.

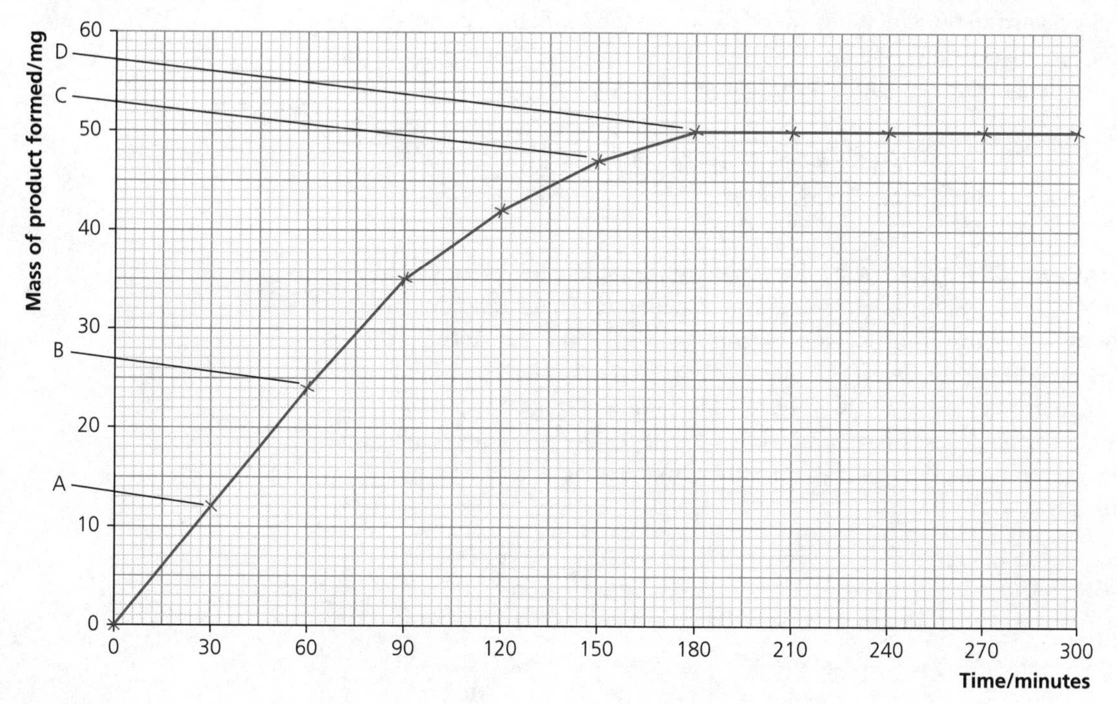

(a) Calculate and derive the units for the rate of reaction between:

A and B

C and D (2 marks)

(b) Explain why these two values are different. (3 marks)

(c) Explain the shape of the curve after point D. (2 marks)

ⓔ Questions asking you to calculate rate are quite common in exams. It is an easy calculation — the rate is found by dividing the mass of product by the time taken to produce it. The units will be whatever the product or substrate is measured in over the unit of time. The rest of the question is fairly straightforward. The main issue with a graph that shows the change in product over time is that students often mix this graph up with a graph that shows the effect of substrate concentration on the rate of reaction.

Student A

(a) Points A and B: 12/30 = 0.4 mg min^{-1}

Points C and D: 3/30 = 0.1 mg min^{-1}

ⓔ **2/2 marks awarded** The calculations and units are correct, for 2 marks.

(b) The rate of reaction was faster at A–B than C–D. This is because C–D is closer to the maximum rate of reaction.

e 0/3 marks awarded Student A has failed to understand that this graph shows a reaction occurring over time, so incorrectly believes that the levelling out of the graph represents a maximum rate of reaction.

(c) After point D all the active sites are full all the time. This means the reaction has reached its maximum rate.

e 0/2 marks awarded Student A's answer reflects what can happen if you fail to identify the type of graph you are being asked to look at. This student clearly learned the shapes of the different rate of reaction graphs and the facts that accompany them (the graph levels out due to all the active sites being full all the time) and then has just reeled it off here. This single mistake has led to student A only scoring 2 marks and underlines the necessity of not just learning the specification content off by heart but really thinking about it and trying to understand it.

Student B

(a) Points A and B: 12/30 = 0.4 mg min^{-1}

Points C and D: 3/30 = 0.1 mg min^{-1}

e 2/2 marks awarded Both rates are calculated correctly and the correct units are given.

(b) The rate of reaction was fastest at A–B as at this point there was a large concentration of substrate so the enzyme concentration was the limiting factor. At C–D the rate of reaction was slower. This is because the concentration of substrate has decreased as it has been converted into products. This means there is a reduced chance of a collision between an active site and a substrate molecule and therefore less product is produced per minute.

e 3/3 marks awarded This is an excellent answer that scores all 3 marks available. There was 1 mark for explaining that A–B was faster due to there being a large concentration of substrate at this point, near the start of the reaction. Student B scores the other 2 marks by identifying that the C–D rate of reaction is slower because some of the substrate has been converted to product so there is less chance of a successful collision between a substrate molecule and an active site.

(c) After point D the graph is flat. This means that the rate of reaction is 0 as all the substrate has been converted into product.

e **2/2 marks awarded** Student B scores both marks by saying the rate of reaction is zero because all the substrates have been converted into products.

Student B's answer illustrates the difference that good understanding and preparation for exams can make. This student has obviously thought carefully about what the graph shows and was able to score all the marks available for this question.

Question 4 Photosynthesis

A student was asked to investigate the effect of light intensity on the rate of photosynthesis. The apparatus available for the investigation are listed below.

- **Lamp**
- **Algae immobilised in alginate beads, supplied in a beaker**
- **Hydrogen carbonate indicator solution (acid — yellow, neutral — red, alkali — purple)**
- **Glass vial**
- **Ruler**
- **Timer**

(a) **Write an outline method for this investigation.** (6 marks)

(b) **What piece of apparatus could be used to give quantitative results in this investigation?** (1 mark)

(c) **Explain why white light is used in this investigation as opposed to shining the light through a colour filter.** (3 marks)

(d) **When the lamp is close to the glass vial containing the algal balls the temperature of the water in the specimen tubes can increase. Explain why this may affect the results of the investigation and give a method of controlling this effect.** (3 marks)

(e) **At high light intensities the indicator turned purple. Explain why this occurred.** (3 marks)

e This question requires you to apply your knowledge of the core practical investigating the rate of photosynthesis to a novel set of apparatus that you probably have not come across before.

Student A

(a) Place the algal balls in a glass vial with the hydrogen carbonate indicator solution.

Shine the lamp on the vial.

Observe the colour change after a set period of time.

e **2/6 marks awarded** This limited answer only scores 2 marks for the set-up of algal balls and indicator, and for observing the colour change after a set period of time. The question asks for an investigation into light intensity, so a fully explained method of changing light intensity is essential.

(b) A colorimeter

e **1/1 mark awarded** This is correct — a colorimeter could give a quantitative value to the colour change.

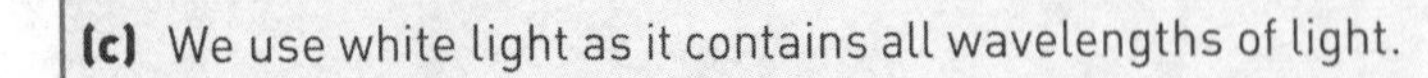

(c) We use white light as it contains all wavelengths of light.

e **1/3 marks awarded** This answer scores 1 mark but needs further development in terms of what the problems of using coloured filters would be.

(d) An increase in the temperature of the water would lead to an increase in the rate of reaction of the enzymes involved in photosynthesis. This could lead to an increase in the rate of photosynthesis, making it difficult to compare the results at different light intensities. A beaker of water could be used as a heat shield to prevent this occurring.

e **3/3 marks awarded** This is an excellent answer, which achieves full marks.

(e) There was an increase in the pH, which caused a change in the colour of the indicator.

e **1/3 marks awarded** This is a good example of how to extract marks from a question you do not fully understand. Student A clearly does not fully understand why the colour changed but has used what they do understand to make a relatively simple statement and get a mark.

Student B

(a) Step 1: Place the algal balls in a glass vial filled with water and hydrogen carbonate indicator.

Step 2: Place the lamp a set distance from the specimen tube.

Step 3: Start the stopwatch and record the time taken for the indicator to change colour.

Step 4: Move the lamp a set distance closer.

Step 5: Replace the used vial with a fresh vial.

Step 6: Repeat the investigation at five or more distances from the specimen tube.

e **6/6 marks awarded** This is an excellent answer, scoring full marks. Setting the plan out as numbered steps helps to organise your answer and is easier for the examiner to mark.

(b) A colorimeter

e **1/1 mark awarded** Correct, for 1 mark.

(c) It is important to use white light as this is made up of all wavelengths of light. Using a coloured filter would limit the wavelengths of light shone on the algal balls. This could then be a limiting factor for photosynthesis.

e **3/3 marks awarded** This is a good answer and scores 3 marks for identifying that white light is made up of all wavelengths of light, that a colour filter would limit the wavelengths reaching the algal balls and that this would therefore limit the rate of photosynthesis.

(d) Increasing the temperature could speed up the rate of photosynthesis.

e **1/3 marks awarded** This answer only scores 1 mark because student B hasn't fully explained why this may affect the results of the investigation or given a way of controlling the temperature change.

(e) At high light intensity the rate of photosynthesis was high. This meant that the plant was taking in a lot of carbon dioxide from the solution. This caused an increase in pH.

e **3/3 marks awarded** This is a good answer that scores all 3 marks.

Question 5 Photosynthetic pigments

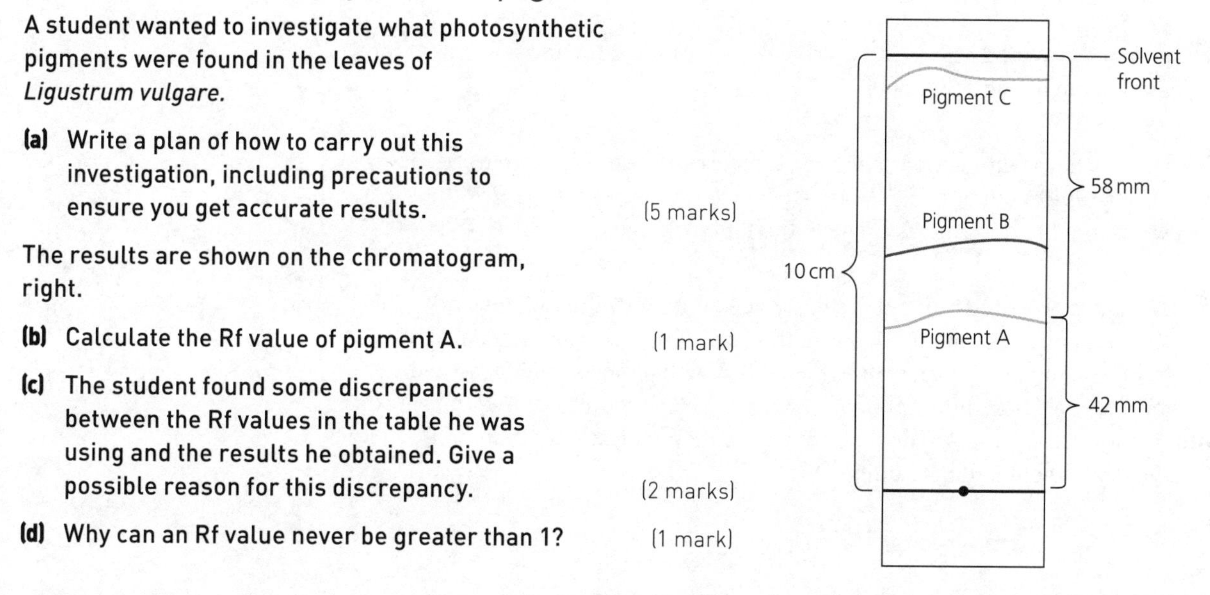

A student wanted to investigate what photosynthetic pigments were found in the leaves of *Ligustrum vulgare*.

(a) **Write a plan of how to carry out this investigation, including precautions to ensure you get accurate results.** (5 marks)

The results are shown on the chromatogram, right.

(b) **Calculate the Rf value of pigment A.** (1 mark)

(c) **The student found some discrepancies between the Rf values in the table he was using and the results he obtained. Give a possible reason for this discrepancy.** (2 marks)

(d) **Why can an Rf value never be greater than 1?** (1 mark)

ⓔ This question requires you to recall the method for one of the core practicals and then apply your knowledge. This underlines the importance of thoroughly learning the methods for the core practicals.

Student A

(a) Step 1: Draw a pencil line near the bottom of a piece of chromatography paper.

Step 2: Grind up some *Ligustrum vulgare* leaves with a pestle and mortar, and add an organic solvent.

Step 3: Put one drop of the extract produced onto the pencil line.

Step 4: Put the chromatography paper into a boiling tube and pour solvent into the tube, ensuring the solvent covers the pigment spot.

Step 5: Wait until the solvent nearly reaches the top of the paper. Take it out of the tube and mark on the paper where the solvent got to.

Step 7: Wait for the chromatogram to dry, and then measure the distance travelled by the pigments. You can then calculate the Rf values.

ⓔ **2/5 marks awarded** This answer does contain some correct points but it is clear that student A has not learned properly the method for this investigation. The main problem with this answer is that two of the points would mean that the investigation would not produce any results. Only placing one drop of extract on the paper would produce a spot that was too light to produce visible pigments. Pouring the solvent so that it covers the spot would cause it to dissolve, meaning that the pigments would not be carried up the paper by the movement of the solvent front.

(b) pigment A Rf value = 42/100 = 0.42

ⓔ **1/1 mark awarded** This is the correct answer, taking the measurement for pigment A from the bottom line.

(c) He may have completed the practical incorrectly, for example used the wrong type of paper and solvent, which would mean that the Rf values would not be the same as those in the table.

ⓔ **1/2 marks awarded** Student A clearly does not know the full answer to this question but has done well to follow up their general 'completed the practical incorrectly' comment, which scores no marks, with a more specific answer. As they have said that different paper and/or solvent would lead to different Rf values, they score 1 mark. The other mark would be for stating that the Rf value is specific to a solvent and medium.

(d) An Rf value is a ratio. To be greater than 1 the pigment would have to travel further than the solvent front. This is impossible.

e 1/1 mark awarded A correct answer, for 1 mark.

Student B

(a) Step 1: Draw a pencil line 25 mm from the bottom of a piece of chromatography paper.

Step 2: Grind up some *Ligustrum vulgare* leaves with a pestle and mortar with an organic solvent such as propanone.

Step 3: Put a small drop of the extract produced onto the centre of the pencil line. Allow it to dry before adding another drop. Repeat this until you have a small, dark pigment spot.

Step 4: Pour the solvent into a boiling tube to a depth of 1 cm.

Step 5: Suspend the chromatography paper in the boiling tube; ensure that the end is in the solvent but that the pencil line and spot remain above the level of the solvent.

Step 6: Observe and wait for the solvent front to nearly reach the top of the paper. Take it out of the tube and mark on the level the solvent front reached.

Step 7: Wait for the chromatogram to dry before measuring the distance travelled by each of the pigments and calculating the Rf values.

e 5/5 marks awarded This is an excellent, thorough answer that scores all 5 marks available.

(b) Pigment A Rf value = 58/100 = 0.58

e 0/1 mark awarded Student B has made an obvious mistake and cost themselves what should have been a straightforward mark. They have measured the distance moved by the pigment incorrectly from the solvent front. This underlines the importance of ensuring that you check back over any measurements or calculations you make in the exam to make sure you have not made any simple mistakes.

(c) Rf values are specific to a type of solvent and medium (paper in this case). The student may have used a different solvent or medium from those that were used to calculate the Rf values in the table, and so got different values.

e **2/2 marks awarded** This is a good answer, which scores both of the marks available.

(d) Rf values are ratios comparing the movement of the pigments with the movement of the solvent front. As the pigments can never travel further than the solvent front, this means that it is impossible for the Rf value to be greater than 1.

e **1/1 mark awarded** This is another good answer, for 1 mark.

Question 6 Respiration

The diagram below shows the apparatus used to measure the rate of respiration in maggots.

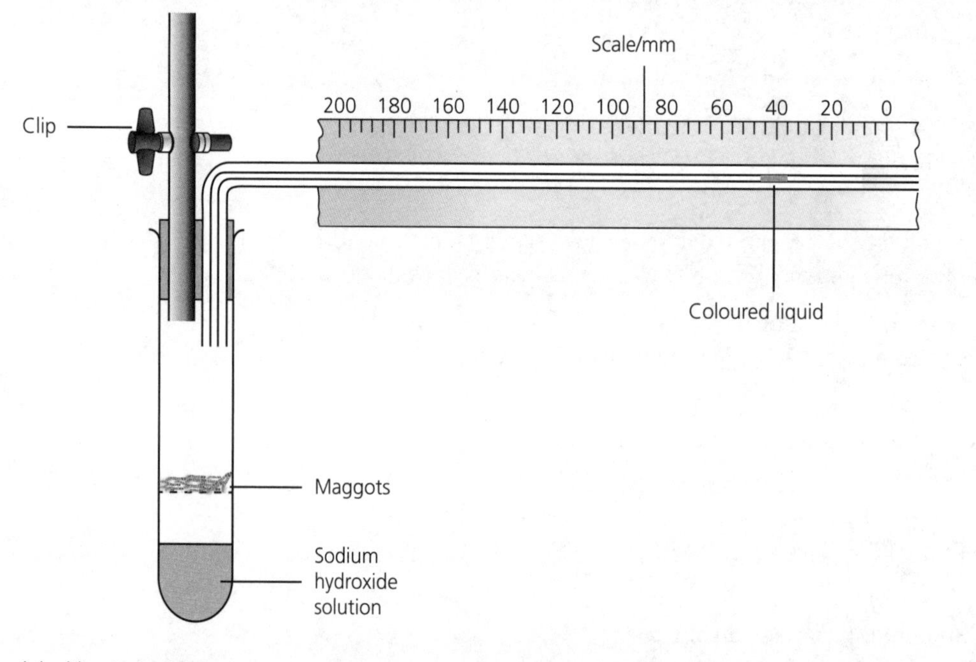

(a) (i) Calculate the rate of oxygen uptake if the coloured liquid moved 32 mm in 0.5 hours. The diameter of the tube containing the coloured liquid is 2 mm. Give the units of your answer. (3 marks)

(ii) When this investigation was repeated different maggots were used that were different sizes from the maggots in the original experiment. What variable should be measured and used in the unit to account for this change? (1 mark)

(iii) The investigation was repeated without the sodium hydroxide. Explain what would happen to the coloured liquid. (2 marks)

(b) The investigation was altered to study anaerobic respiration in yeast cells. The new apparatus is shown opposite.

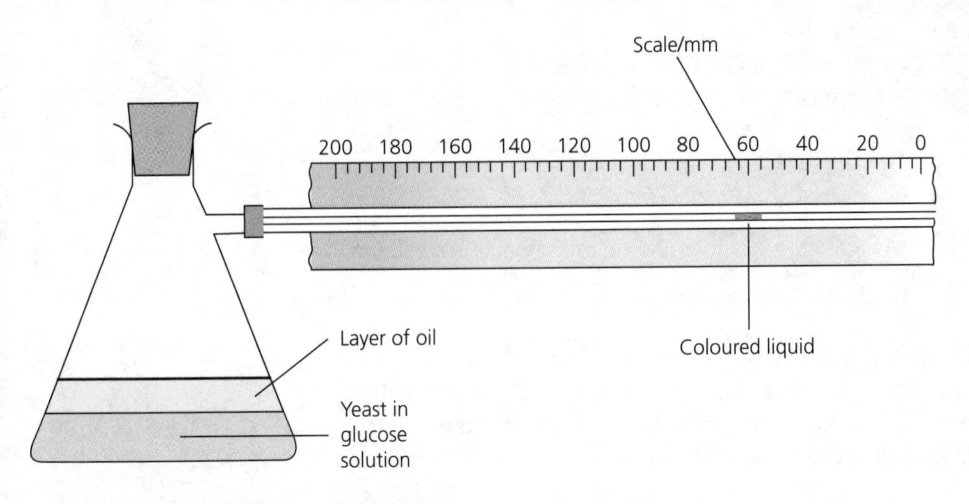

(i) Why is the oil layer present? (1 mark)

(ii) Predict the behaviour of the coloured liquid and explain your answer. (3 marks)

(iii) Would this second investigation produce different results if animal tissue was used instead of yeast? Explain your answer. (2 marks)

e This question requires you to understand the fundamental biology underlying the respirometer core practical in order to answer questions on the standard set-up and a further investigation. This underlines the importance of understanding *why* things happen as opposed to just learning the expected results of the core practicals.

Student A

(a) (i) volume of oxygen = πr^2 × distance moved by liquid

volume = $\pi \times 1^2 \times 32$

= 101 mm^3

rate = volume of oxygen/time

= 101/0.5 = 202 mm

e **2/3 marks awarded** The student has done all the hard work and got the correct numerical answer but has not used the correct unit so loses 1 mark. Rate of oxygen uptake requires a volume component and a time component.

(ii) Mass of maggot

e **1/1 mark awarded** A correct answer, for 1 mark. Finding the mass of the maggots and incorporating it into the units of the answer would help to control the variability caused by using different-sized maggots.

(iii) The coloured liquid would remain stationary.

e **1/2 marks awarded** This is correct and scores 1 mark but there is no explanation of why the coloured liquid did not move (the volume of gas would not change so the pressure would not change either).

(b) (i) The oil layer prevents any oxygen getting to the yeast suspension which ensures only anaerobic respiration is occurring.

e 1/1 mark awarded A correct answer, scoring 1 mark.

(ii) Carbon dioxide is released by the yeast cell, causing the coloured liquid to move to the right.

e 1/3 marks awarded While this answer is correct it only scores 1 of the 3 marks available (for stating that carbon dioxide is produced). To score the other 2 marks student A had to state why the carbon dioxide is produced and that it is the change in pressure that causes the move to the right.

(iii) The coloured liquid would move to the left, as there is no carbon dioxide produced by the animal tissue in anaerobic respiration.

e 1/2 marks awarded This answer scores 1 mark for stating that the animal tissue will not produce any carbon dioxide, but then loses the second mark for incorrectly saying the liquid would move. If no gas is produced there will be no change in pressure, which would mean the liquid will not move.

Student B

(a) (i) volume of oxygen = $\pi r^2 \times$ distance moved by liquid

$r = 1\,\text{mm}$

volume of oxygen = $\pi \times 1^2 \times 32$

$= 101\,\text{mm}^3$

rate = volume of oxygen/time

$= 101/0.5$

$= 202\,\text{mm}^3\,\text{hour}^{-1}$

e 3/3 marks awarded Student B has correctly calculated the answer and derived the units, so scores all 3 marks.

(ii) Mass of the maggots

e 1/1 mark awarded Correct for 1 mark.

(iii) The coloured liquid would not move as the volume of oxygen taken in would be roughly equal to the volume of carbon dioxide released, so the pressure in the tube would not change.

e 2/2 marks awarded This the correct answer and scores both available marks.

(b) (i) The oil layer prevents carbon dioxide leaving the solution.

0/1 mark awarded This is an incorrect answer — the oil layer is present to prevent oxygen reaching the yeast.

(ii) The yeast cells would produce carbon dioxide during anaerobic respiration, due to the decarboxylation of pyruvate to form ethanal. This would increase the pressure and lead to the coloured liquid moving to the right.

3/3 marks awarded This is a good answer, which links the expected experimental results to the theory of anaerobic respiration.

(iii) The coloured liquid would not move if the investigation was repeated with animal tissue. This is because no gas is produced by anaerobic respiration in animal tissue so there would be no change in pressure.

2/2 marks awarded This is a good answer, for full marks.

Question 7 Investigating the effect of gibberellin

A student was investigating the effect of gibberellin concentration on the production of enzyme by a seed. She was given a $4\,g\,dm^{-3}$ solution of gibberellin.

(a) The student ensured that the embryos were removed from the seeds used. Explain the importance of this. (2 marks)

She carried out a serial dilution of the initial $4\,g\,dm^{-3}$ gibberellin solution, as shown in the diagram below.

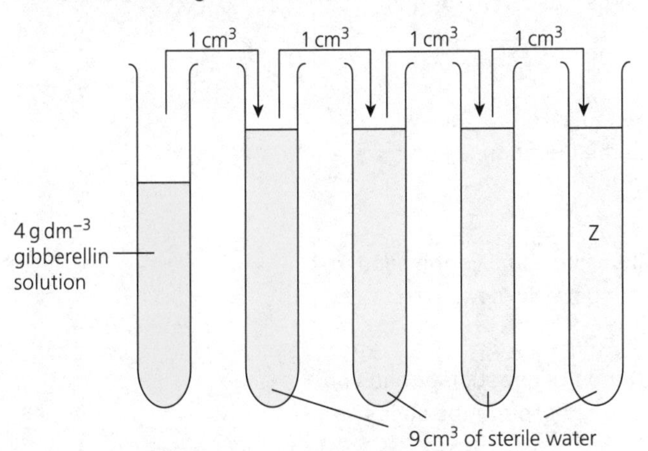

(b) (i) What was the concentration of the solution in tube Z? (2 marks)

(ii) The student wanted to produce a $10\,cm^3$ solution with a concentration of $8 \times 10^{-3}\,g\,dm^{-3}$. Explain how they could use a new serial dilution to achieve this. (2 marks)

The seeds were soaked in the gibberellin solutions and then placed onto an agar plate. After a period of time the agar plate was flooded with iodine. Most of the agar turned a blue/black colour, while a circular area around each of the seeds remained the original colour.

(c) Explain this result. (3 marks)

(d) The diameter of the circular area around a seed that was treated with the solution from tube Z was 11 mm. Calculate the area of this circle. (1 mark)

e This question involves some fairly complicated maths. It is important to have thoroughly practised all the required maths skills and be comfortable in using them all in a variety of situations.

Student A

(a) As the concentration of gibberellin is the independent variable it is important the concentration is known. If the embryo was not removed the concentration would not be definitely known.

e **1/2 marks awarded** This answer scores 1 mark for explaining the importance of ensuring that the concentration of gibberellin is known and that this would not be possible if the embryo was present. Student A has not stated that it is the embryo that would be releasing the gibberellin and therefore causing the concentration to be unknown, so does not score the second mark.

(b) (i) The dilution factor is 1×10^{-4}.

e **1/2 marks awarded** Student A has correctly identified the dilution factor but has not stated what concentration this gives.

(ii) Dilute the solution three times to 1×10^{-3} and then use 8 cm^3 of this solution.

e **0/2 marks awarded** If student A had read the question carefully they would have seen that this answer must be incorrect because the question asked for a final solution of 10 cm^3. Student A scores no marks for this part.

(c) The starch in the agar turns iodine blue/black. Gibberellin causes the seed to release amylase. The amylase diffuses into the agar and breaks down the starch.

e **2/3 marks awarded** Student A has clearly understood the question but has not developed the answer fully and made reference to the areas around the seeds, which do not change colour.

(d) radius = 5.5 mm

area = $\pi \times 5.5^2$

= 95 mm^2

e 1/1 marks awarded This is the correct answer and scores the available mark. However, it is good practice to write out working fully.

Student B

(a) The embryo of the seeds must be removed as it produces gibberellin. If the embryo was left it would produce gibberellin, meaning that the concentration on each of the seeds would no longer be known.

e 2/2 marks awarded This is an excellent answer, which scores both of the available marks.

(b) (i) The dilution factor is 1×10^{-4}. This means that the concentration of the solution is 4×10^{-4} g dm^{-3}.

e 2/2 marks awarded Student B has given the dilution factor and the concentration, so scores both marks.

(ii) By diluting 2 cm^3 of the original solution into 8 cm^3 of sterile water. This would give a concentration of 8×10^{-1} g dm^{-3}. This solution could then be diluted a factor of 10 twice (1 cm^3 of solution into 9 cm^3 of distilled water each time). This would give a 10 cm^3 solution of concentration 8×10^{-3} g dm^{-3}.

e 2/2 marks awarded There are several different ways of getting the correct concentration. This is an efficient, well-explained method, which scores both marks.

(c) The agar contains starch, which turns iodine blue/black. The gibberellin causes the seed to release amylase. This amylase diffuses into the agar and hydrolyses the starch. The areas where this occurs would no longer contain starch, and so remain the original colour.

e 3/3 marks awarded This is a good answer, which fully explains the results, and so gains all 3 marks.

(d) radius = diameter/2

radius = 11/2

= 5.5 mm

$$\text{area} = \pi r^2$$
$$= \pi \times 5.5^2$$
$$= 95\,\text{mm}^2$$

e **1/1 mark awarded** This answer contains detailed working and the correct answer.

Question 8 Transpiration

The apparatus below was used in an investigation.

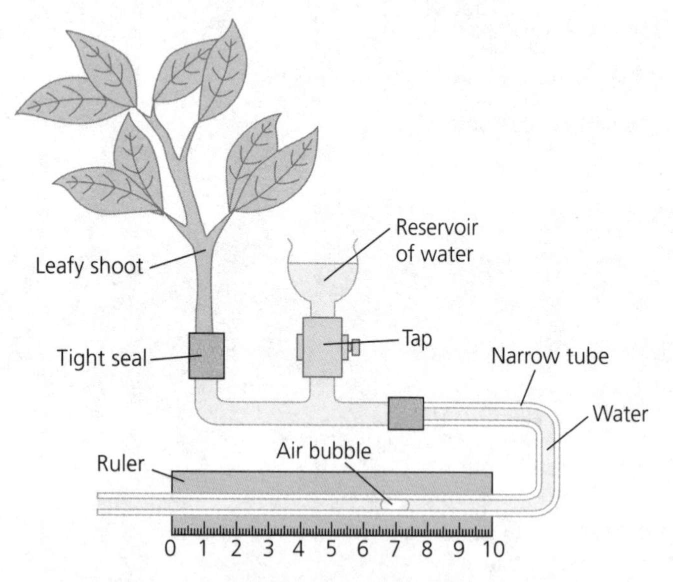

(a) **Why is it incorrect to state that the dependent variable in the investigation is the rate of transpiration?** (2 marks)

(b) **(i)** **The independent variable of this investigation was temperature. At 20°C the bubble moved a distance of 5 cm in a tube of diameter 1 mm in one and a half hours. Calculate the rate of water uptake in $cm^3\,min^{-1}$.** (2 marks)

(ii) **The total surface area of the underside of the leaves on the shoot was 30 cm^2. Use this information to give the rate of water uptake per unit area of the underside of the leaves.** (1 mark)

In a follow-up investigation, temperature was kept constant, and a plastic bag was placed completely over the shoot and all the leaves. The rate of water uptake was then measured.

(c) **State the independent variable of this investigation and predict how the rate of water uptake would compare to the same shoot without a plastic bag placed over it.** (3 marks)

e The last part of the question asks you to predict the results of an investigation. This underlines the importance of not only knowing how to carry out the core practicals but also what the expected results are, and why.

Student A

(a) Potometers measure the uptake of water.

e 0/2 marks awarded This answer scores no marks. It does not state that the rate of transpiration is not equal to the rate of water uptake, or explain why.

(b) (i) $\pi r^2 = 0.008\,cm^2$

volume = 0.008×5

$= 0.04\,cm^3$

e 1/2 marks awarded Student A has not fully written out their working and in their rush to complete the question has missed out the last part. They have calculated the volume of water taken up but then forgotten the final step to get the rate.

(ii) $0.04/30 = 1.3 \times 10^{-3}\,cm^2$

e 0/1 mark awarded Student A has correctly realised that they need to divide their answer to (b)(i) by the total area, but as that answer was wrong this answer is also wrong. If student A had taken the time to look back over their answers they would have realised that the answers to (b)(i) and (ii) are both incorrect, as neither of them is actually a rate.

(c) The plastic bag traps humid air near the stomata, reducing the diffusion gradient between the inside of the leaf and the outer air. This leads to an increase in the rate of transpiration and therefore an increase in water uptake.

e 1/3 marks awarded This answer only scores 1 mark for stating that the increase in humidity decreases the diffusion gradient between the air and the inside of the leaf. Student A has failed to state that humidity is the independent variable, even though it specifically asks for this in the question. The answer then goes on to say, incorrectly, that a higher humidity leads to an increase in the rate of transpiration. Humidity is one of the only factors that causes a *decrease* in the rate of transpiration when it increases and therefore also a decrease in the rate of water uptake.

Student B

(a) This is a potometer, which measures the rate of water uptake. While most of the water taken in by the plant is lost in transpiration some is used in photosynthesis and other processes.

e 2/2 marks awarded This is the correct answer and scores both marks for correctly identifying the dependent variable as the rate of water uptake and explaining why this does not equal the rate of transpiration.

(b) (i) volume of water = $\pi r^2 \times$ distance travelled

$r = 0.1/2 = 0.05$

$\pi \times 0.05^2 = 0.008\,cm^2$

volume = 0.008×5

$= 0.04\,cm^3$

rate = $0.04/90$

$= 4 \times 10^{-4}\,cm^3\,min^{-1}$

e **2/2 marks awarded** This is the correct calculation of volume of water taken in and then rate, so scores both available marks.

(ii) $(4 \times 10^{-4})/30$

$= 1.5 \times 10^{-5}\,cm^3\,min^{-1}\,cm^{-2}$

e **1/1 mark awarded** Student B has correctly calculated the rate per area of leaf and also used the appropriate units.

(c) The independent variable is the humidity of the air. Placing the plastic bag over the shoot traps water vapour that has evaporated from the interior surfaces of the leaf and diffused through the stomata. This increases the humidity of the air around the leaves, therefore decreasing the rate of water uptake. This is because there is a reduced concentration gradient between the inside of the leaf and the air around the stomata, lowering the rate of diffusion of water through the stomata.

e **3/3 marks awarded** This is an excellent answer, which correctly identifies humidity as the independent variable and then explains in detail the effect of an increase in humidity on the rate of water uptake, scoring all 3 marks.

Question 9 Counting bacteria

Researchers were studying the number of bacteria over a period of time in a small pond. They used the turbidity of the pond water to determine the changes in the bacterial population over time.

(a) Evaluate this method of counting for this purpose. (4 marks)

(b) The researchers found that the absorbance at the start of the bacteria's log phase was 0.61 and at the end of the log phase it was 1.89. This change took 7 hours. Use this information and the equation below to calculate the exponential growth constant of the bacteria during the log phase.

$$k = \frac{\log_{10} OD_1 - \log_{10} OD_0}{\log_{10} 2 \times t}$$

where *k* is the exponential growth rate constant, OD_0 is the absorbance at the start of the log phase, OD_1 is the absorbance reading after time *t* (using a reading towards the end of the log phase) and *t* is the time the culture has been growing. (2 marks)

ⓔ The first part of this question requires you to apply your practical knowledge to a specific example. Make sure that you read the question carefully and tailor your answer appropriately rather than just writing general points. The second part of the question covers logs. Students often lack confidence in using logs but they are actually very straightforward — just make sure that you know how to do them on your calculator.

Student A

(a) Measuring the turbidity of the water is a quick and simple method of estimating the number of bacteria in a sample. It can also be done in the field (at the side of the pond) using a portable colorimeter, so there is no need to return the samples to the lab.

It is not a viable count method, however, because it counts both living and dead bacteria, possibly leading to overestimation. It is also not very accurate as anything that causes the pond water to have a higher absorbance (not just bacteria) will affect the results.

ⓔ **4/4 marks awarded** This is a good answer, which scores all 4 marks available, giving two advantages and two disadvantages that are specific to the example in the question.

(b) = 0.491/1.15

= 0.427

ⓔ **0/2 marks awarded** This is the incorrect answer and, because student A has not provided any real working, it is impossible for them to gain the other mark. It appears that they multiplied 2 and 7 and then found the log of this answer. That one careless mistake has cost them 2 marks and highlights the importance of checking all answers, particularly numerical ones, carefully, and providing detailed working so that if a final answer is wrong there may still be the possibility of picking up a mark.

Student B

(a) Measuring turbidity is not a viable count method so can lead to the overestimation of the number of bacteria present because it includes both dead and living bacteria. It is also not an accurate method as it relies on the change in light absorbance by the solution being due to the bacteria present and no other factor.

e **2/4 marks awarded** This answer gives two well-explained disadvantages, but as this is a 4-mark question, more is clearly required. There are advantages in measuring turbidity, particularly in the field, as in this example.

(b) $k = (\log_{10} OD_1 - \log_{10} OD_0)/(\log_{10} 2 \times t)$

$= (\log_{10} 1.89 - \log_{10} 0.61)/(\log_{10} 2 \times 7)$

$= (0.276 - 0.215)/2.11$

$= 0.491/2.11$

$= 0.233$

e **2/2 marks awarded** This is a well-set-out calculation, which scores both available marks.

Question 10 Cell division

The table below shows the results of an investigation into cell division in garlic root tip cells.

Phase	Number of cells
Interphase	96
Prophase	25
Metaphase	9
Anaphase	5
Telophase	16

(a) **Calculate the mitotic index of this sample.** (1 mark)

(b) **What can be inferred about the length of the stages of the cell cycle from these results? Explain your answer.** (3 marks)

(c) **A student would like to carry out a further investigation into whether nitrate deficiency or phosphate deficiency has the greatest effect on cell division in garlic root tips. Explain how she could do this. She is provided with a growth medium lacking in nitrates, a growth medium lacking in phosphates and a complete growth medium that is not lacking in any minerals.** (5 marks)

e This question starts with some data analysis and then moves on to a general experimental design question with a high mark allocation. With these questions be as detailed as possible and cover each aspect of good experimental design.

Student A

(a) number of cells = 151

mitotic index = 55/151

= 0.36

e **1/1 mark awarded** Correct, scoring 1 mark.

(b) Anaphase is the fastest phase as it has the lowest number of cells in it.

e **1/3 marks awarded** This is correct, for 1 mark, but this is a 3-mark question, so significantly more detail is required. Student A should have compared interphase with the other stages, and interphase with mitosis as a whole.

(c) The student should set up an experiment with a garlic bulb in nitrate-depleted growth medium, another in phosphate-depleted growth medium and a control bulb in complete growth medium. They should sample a root tip of each every 2 hours, count the number of cells undergoing mitosis and then calculate the mitotic index of each bulb. They should repeat this every 2 hours over a day. Once they have all their results, they can then compare them and see if the different media had any effect on the mitotic indexes of the bulbs.

e **2/5 marks awarded** Student A does pick up 2 marks for this answer, one for setting up the bulbs including a control and the other for saying that they would count the cells undergoing mitosis, calculate the mitotic index and use this to compare the effect of nitrate and phosphate deficiency. However, the rest of the answer is poor. They make no mention of a hypothesis or replicates. The time scale is unrealistic as it is unlikely that you would see much effect after only a few hours. They have also failed to state that if their results were suitable, they would use a statistical test to determine if their results showed significant differences. This highlights the importance of writing detailed, well-thought-out answers that cover all parts of an investigation when answering high-mark questions such as this.

Student B

(a) 55/96

= 0.57

e **0/1 mark awarded** Student B has calculated the mitotic index incorrectly. They have divided the number of cells in mitosis by the number of cells in interphase rather than the total number of cells. Carelessness has cost them a straightforward mark.

(b) Interphase is the longest stage of the cell cycle and each cell spends more time in interphase than undergoing mitosis as a whole. You can see this as the total number of cells in all four stages of mitosis is still less than the number of cells in interphase. The fastest stage of mitosis is anaphase because the least number of cells (5) are in it. The longest stage of mitosis is prophase, because it has the most cells in it after interphase.

e **3/3 marks awarded** This is a thorough answer that fully explains what the different number of cells in each stage suggests about the length of the stages. It is particularly good that student B has compared the number of cells in interphase with the total number of cells in mitosis rather than just the individual stages. This has allowed them to make the statement that each cell spends longer in interphase than in the whole of mitosis.

(c) The independent variable in this investigation is the mineral the garlic bulb is deficient in. A good dependent variable to use would be the mitotic index. First she should write a hypothesis, possibly something like: 'There will be a significant difference between mitotic index in the garlic growing in a nitrate-deficient solution and the garlic growing in a phosphate-deficient solution'. She should then set up a number of garlic bulbs with their roots in mineral solutions that are deficient in nitrates and a number of bulbs with roots in solutions that are deficient in phosphate. She should also set up bulbs growing in a complete mineral solution to act as a control.

After a week she should then sample a root tip from each of the bulbs and count the number of cells undergoing mitosis and use this to calculate a mitotic index. She should repeat this process over a number of weeks. Once she has a good range of data, and if the results are suitable, she should use an appropriate statistical test to determine if there is a significant difference between the mitotic indexes of the different garlic bulbs.

e **5/5 marks awarded** This is a detailed answer that scores full marks. Student B has given a hypothesis and explained how they would collect the data, including a reasonable timescale, planning to do replicates and having a control. Finally they have stated that they will use a statistical test to see if any differences they find are significant.

Index